Kindergarten

MATH LITERCY
workbook

Math Literacy : Kindergarten Workbook Mix Game
Learning By Playing

Copyright: Published in the United States by Mj.Mitzery
Published August 2018

All rights reserved. No part of this publication may be reproduced, stored in retrieval system, copied in any form or by any means, electronic, mechanical, photocopying, recording or otherwise transmitted without written permission from the publisher. Please do not participate in or encourage piracy of this material in any way. You must not circulate this book in any format Mj. Mitzery does not control or direct users' actions and is not responsible for the information or content shared, harm and/or actions of the book readers.

ISBN-13: 978-1726216708

ISBN-10: 1726216705

I will practice

Tracing Number

Count and Match

Compareing

Addition

Name _____

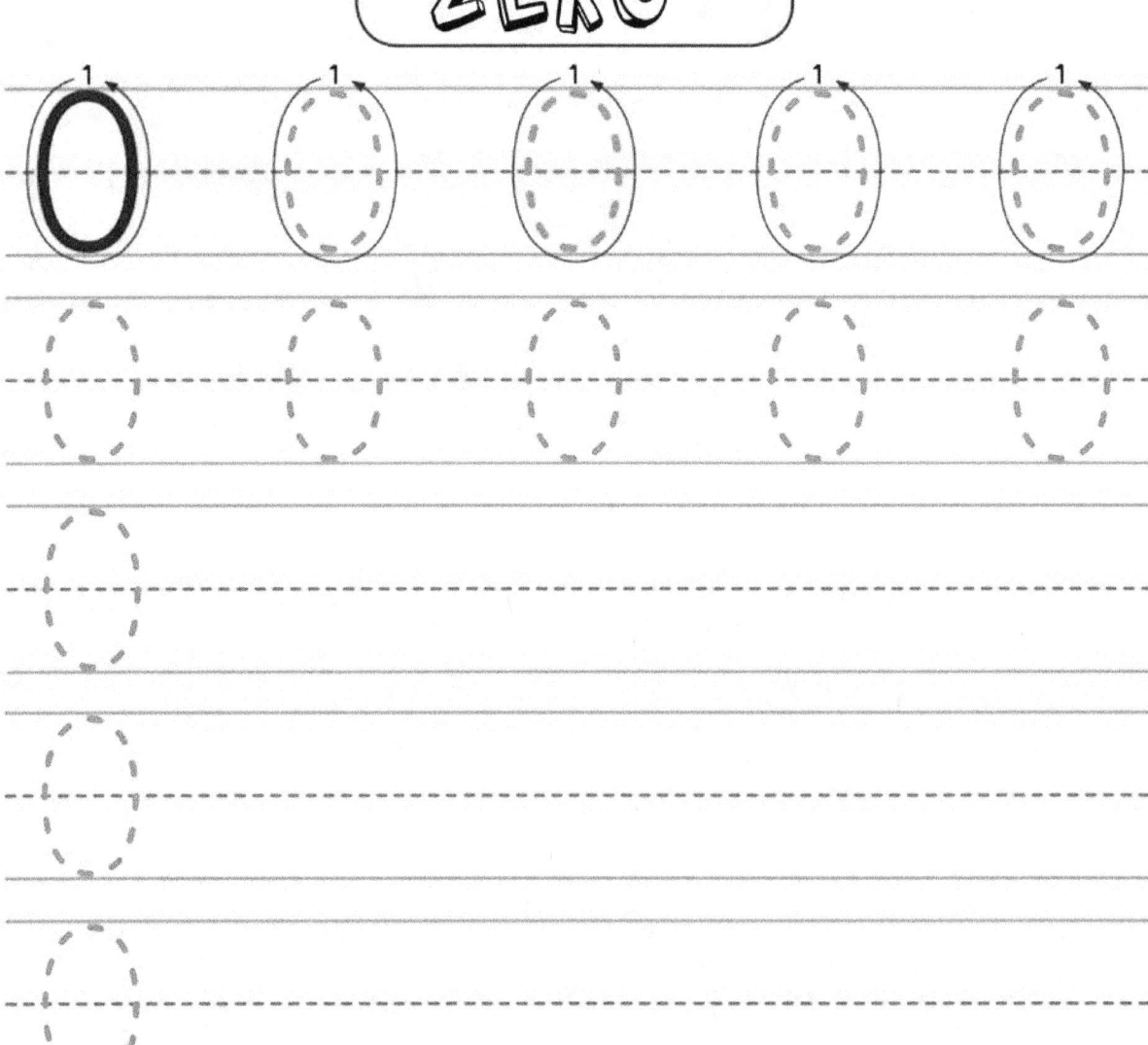

Name _____

Name _____

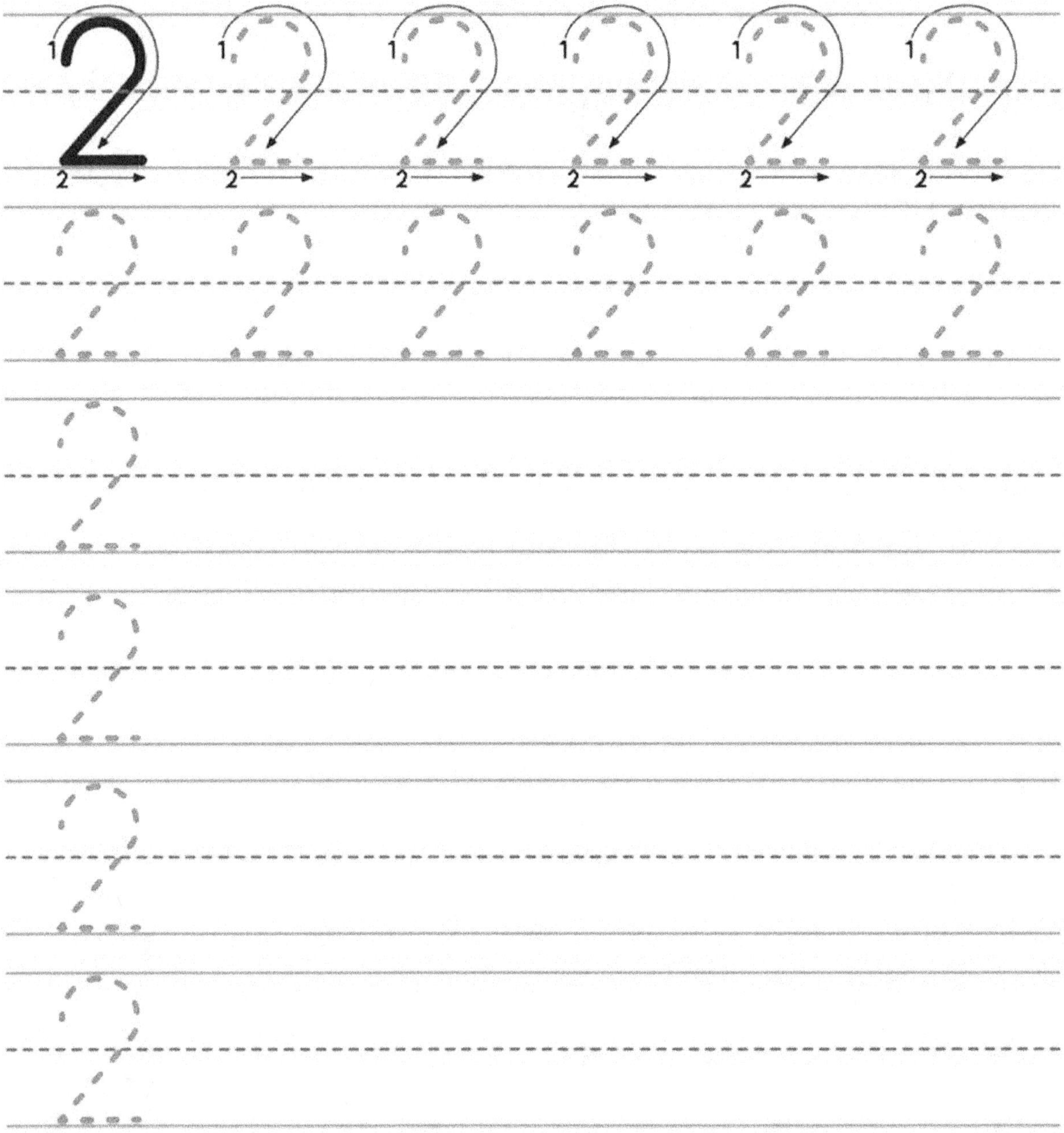

Name _____

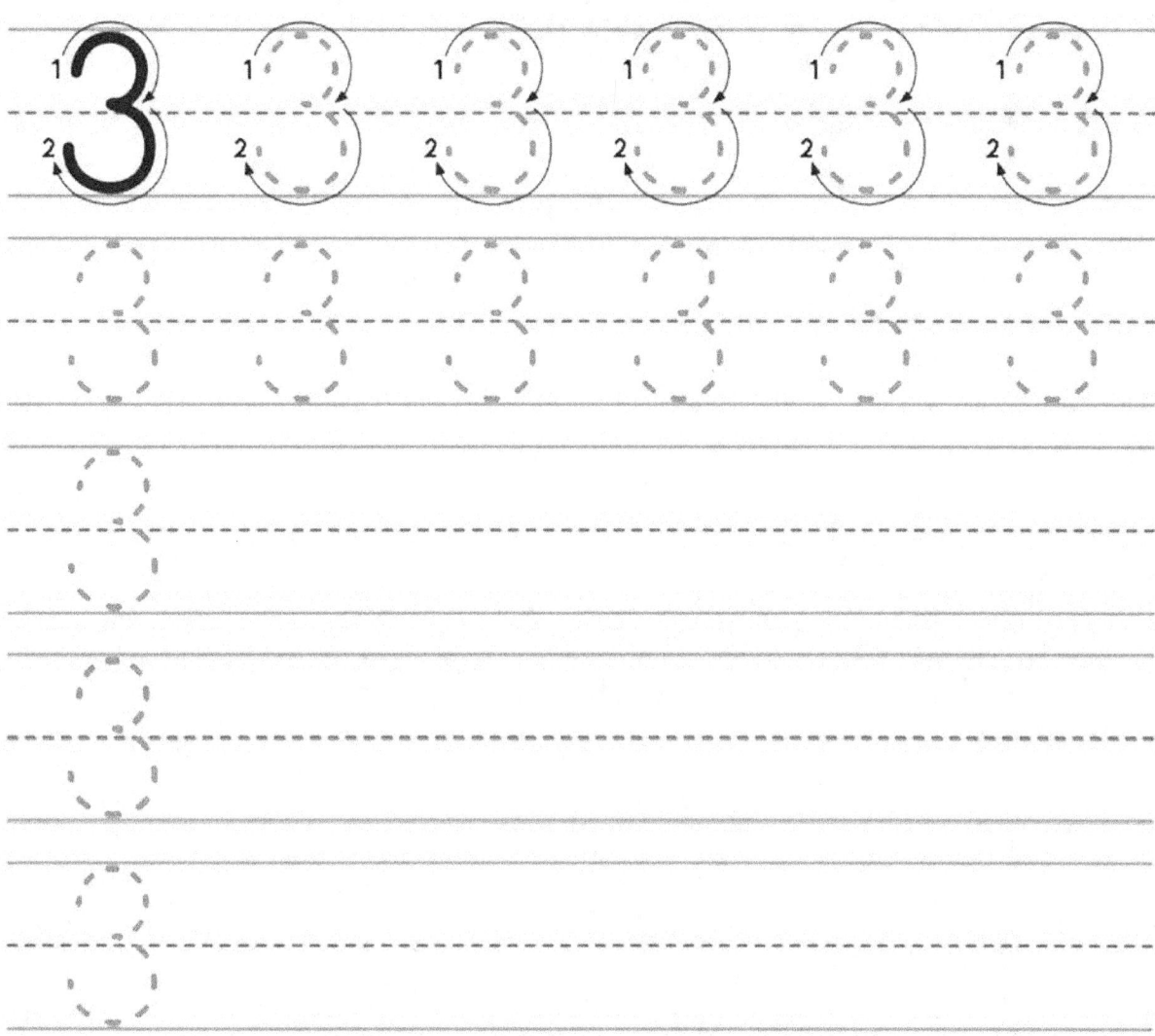

Name

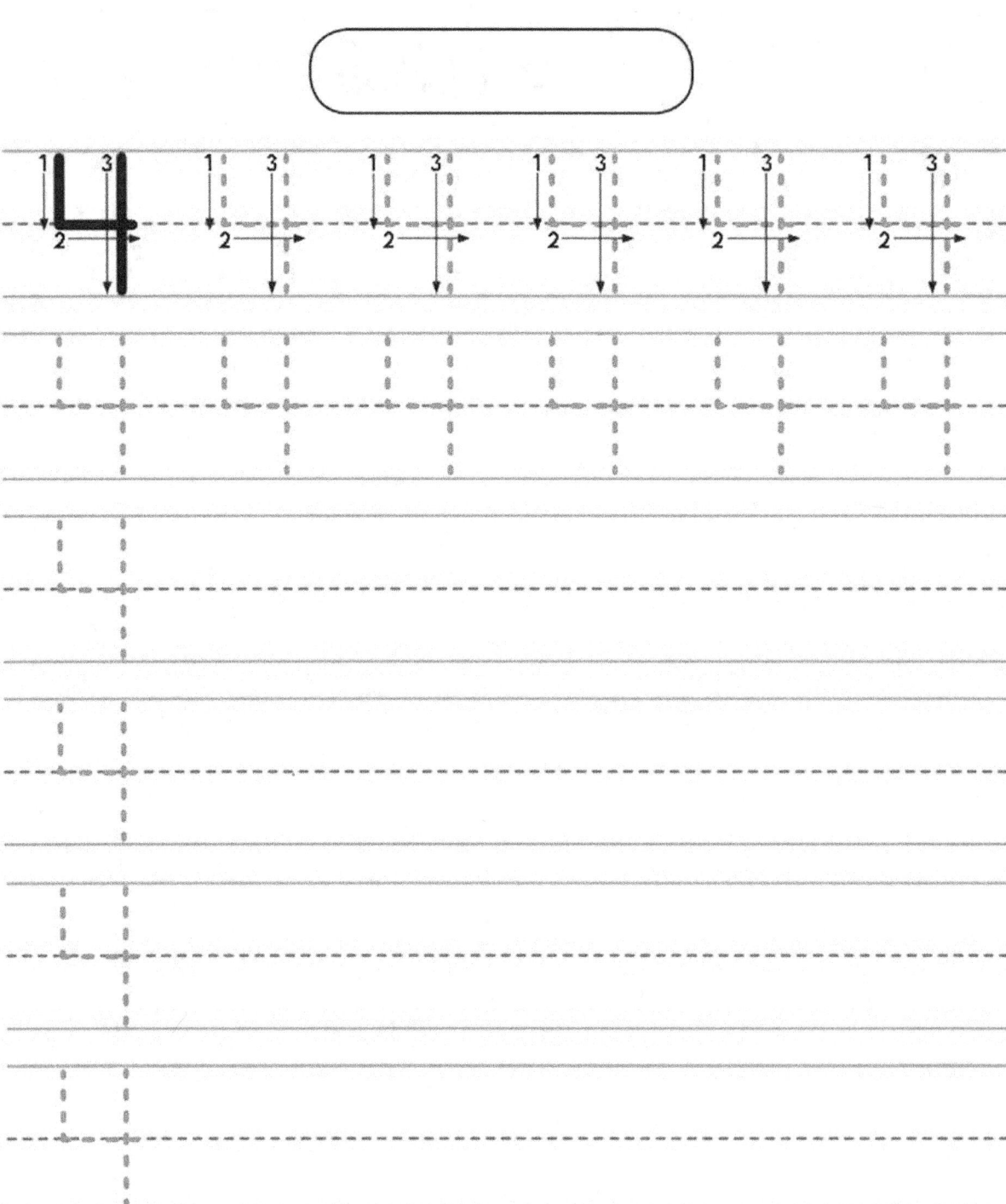

Name

FIVE

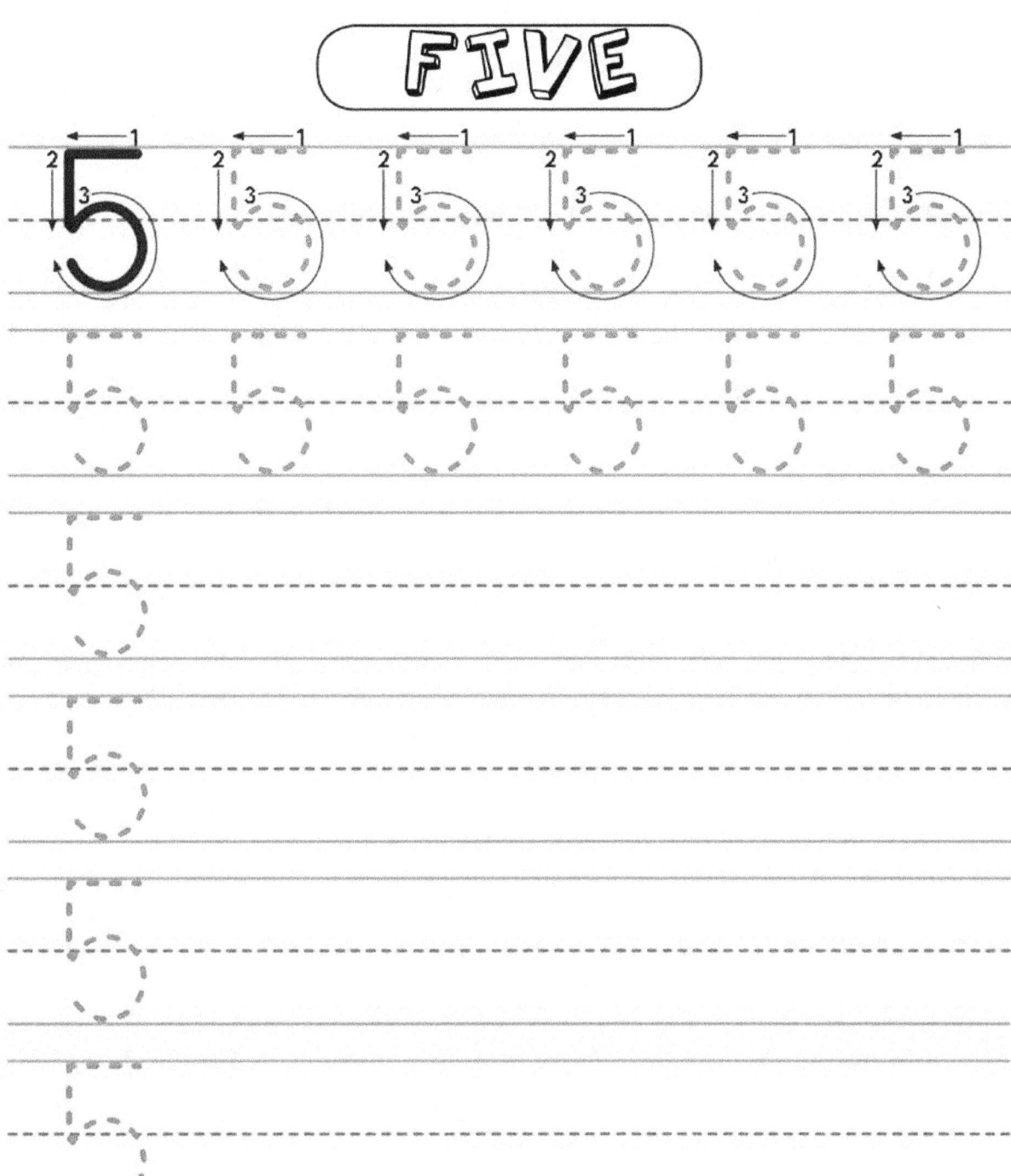

Name _____

6 6 6 6 6 6 6

Name _____

SEVEN

7

Name _____

EIGHT

8 8 8 8 8 8

Name _____

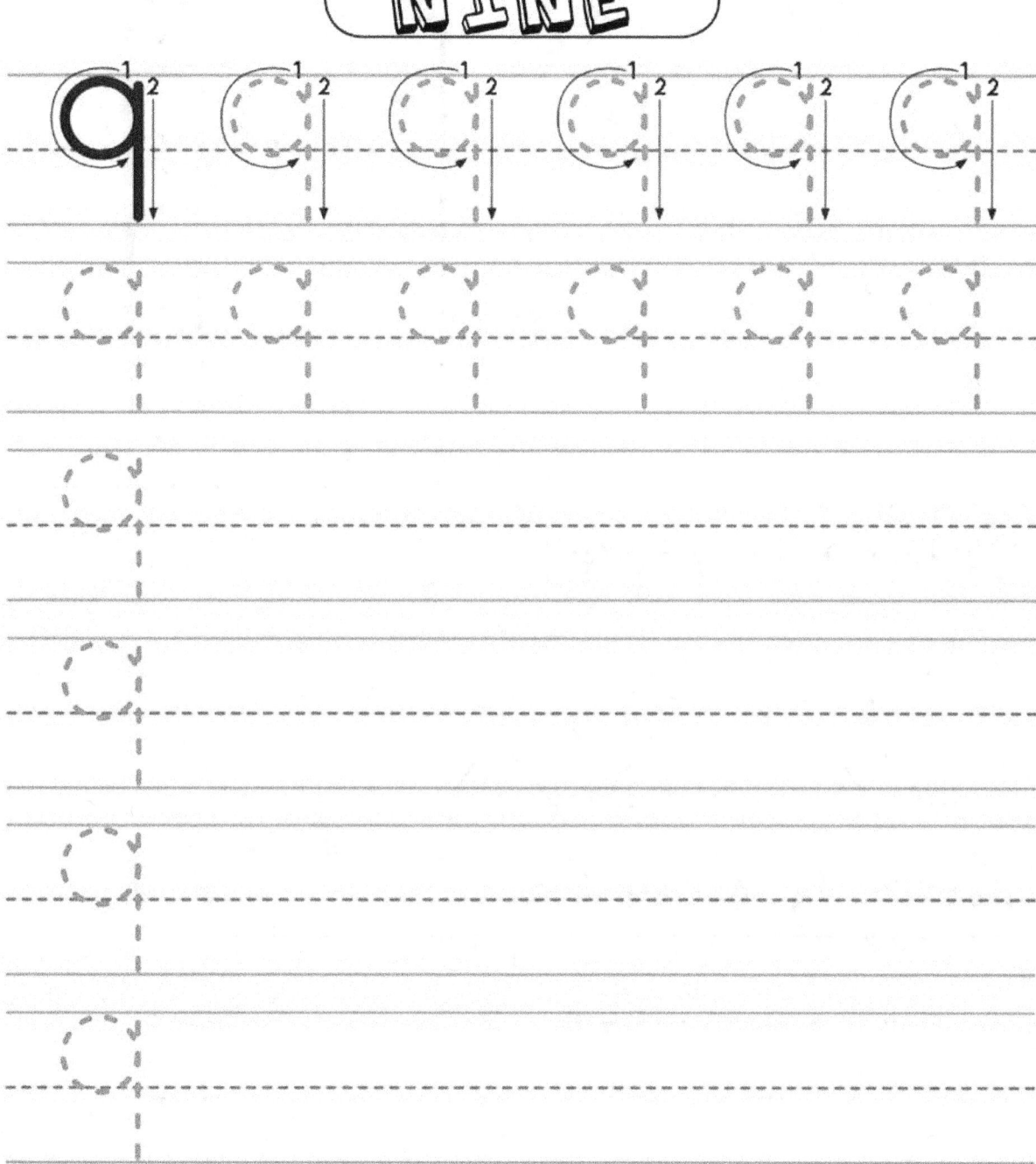

Circle Number 1s

Trace

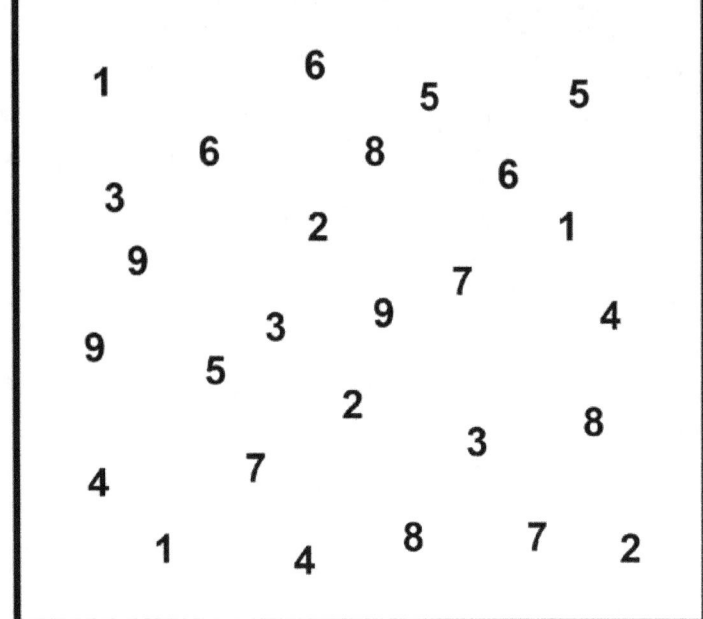
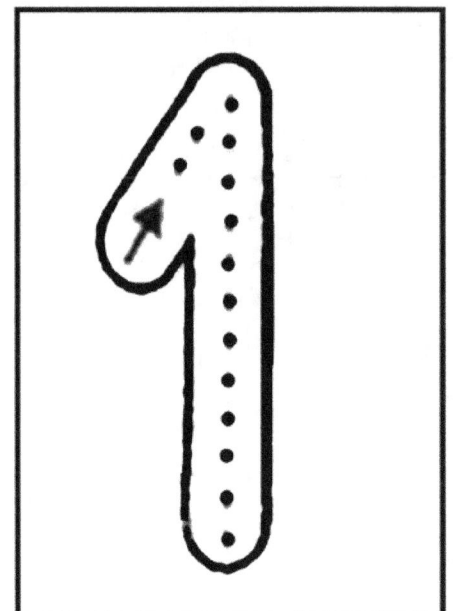

Color 1 Triangle

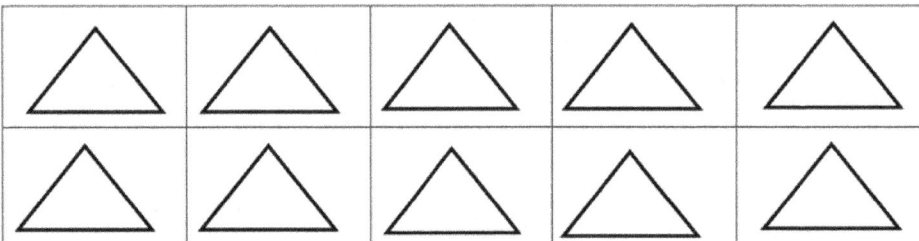

Tracing Number 1s

Circle Number 2s

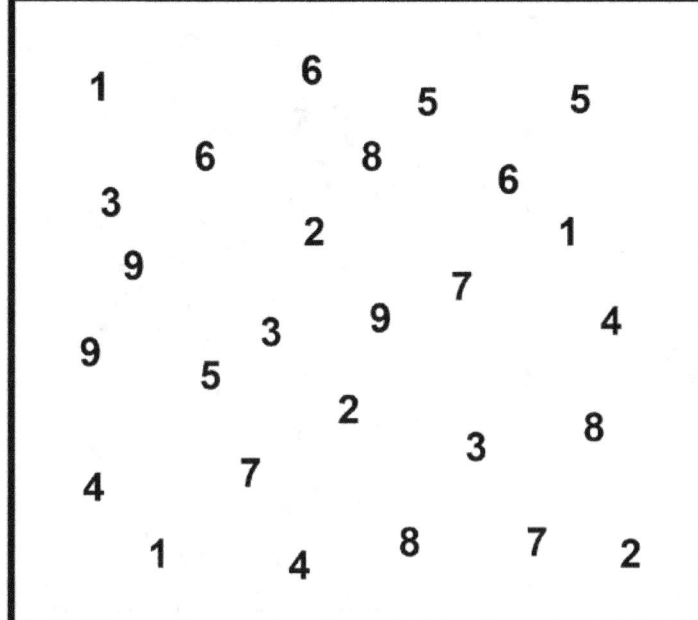

Trace

Color 2 Triangles

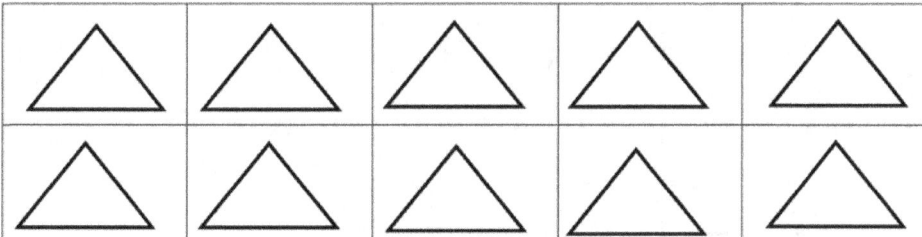

Tracing Number 2s

2 -- 2 -- 2 -- 2 -- 2

2 -- 2 -- 2 -- 2 -- 2

Circle Number 3s

Trace

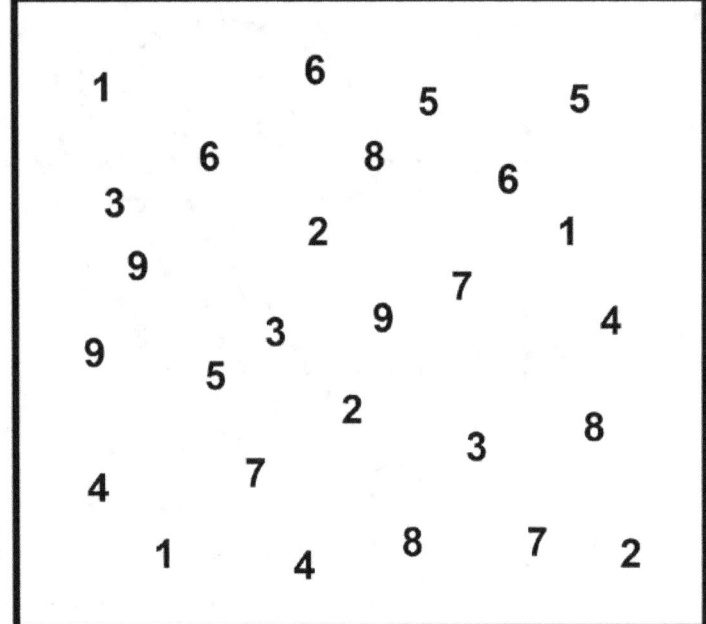

Color 3 Triangles

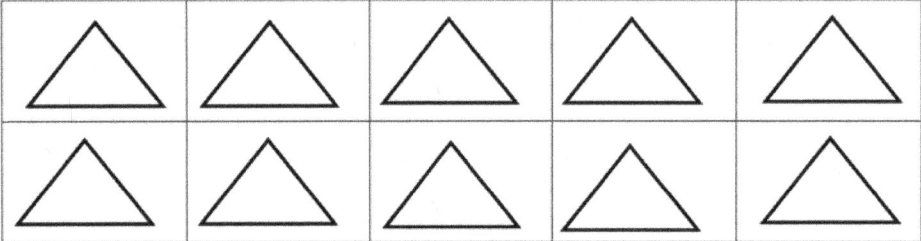

Tracing Number 3s

Circle Number 4s

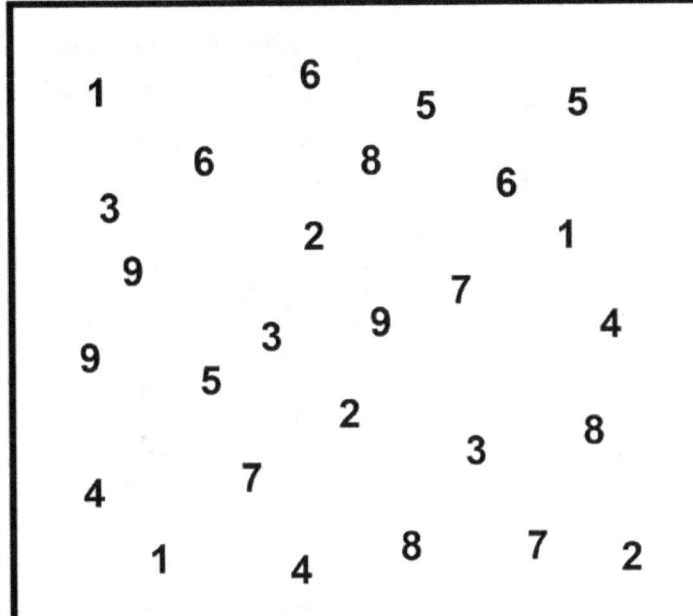

Trace

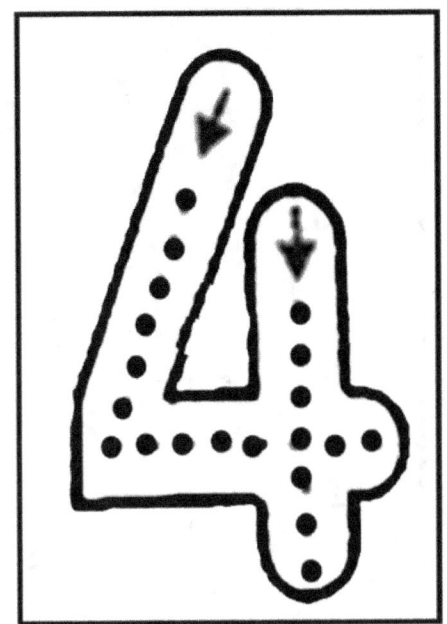

Color 4 Triangles

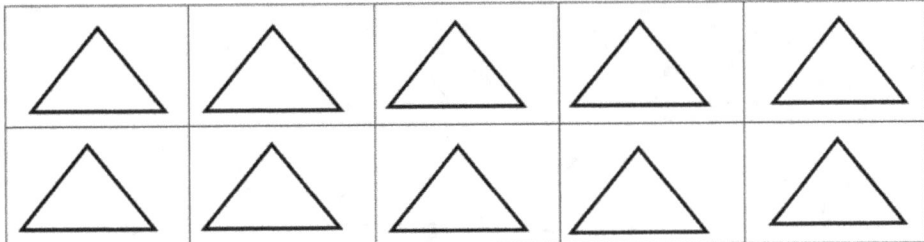

Tracing Number 4s

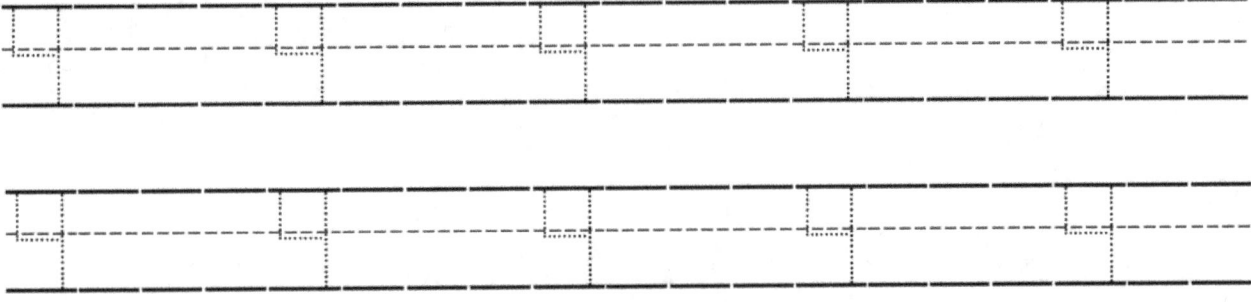

Circle Number 5s
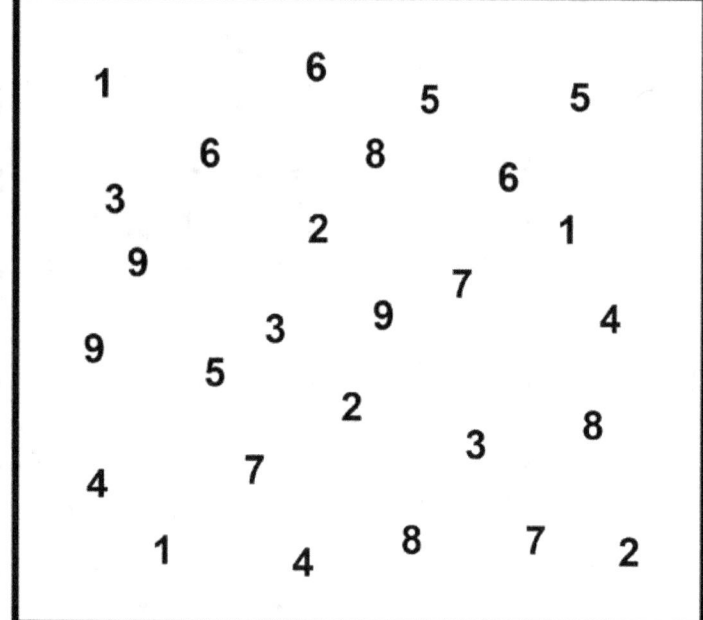

Trace

Color 5 Triangles
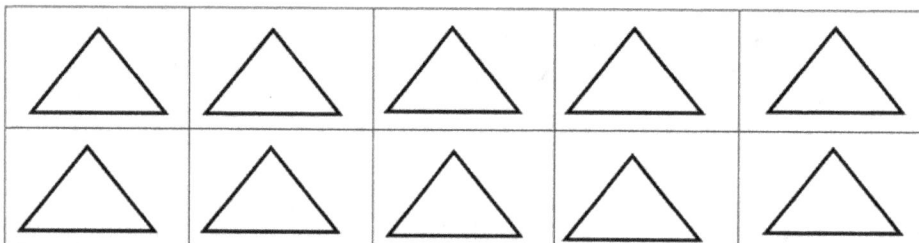

Tracing Number 5s
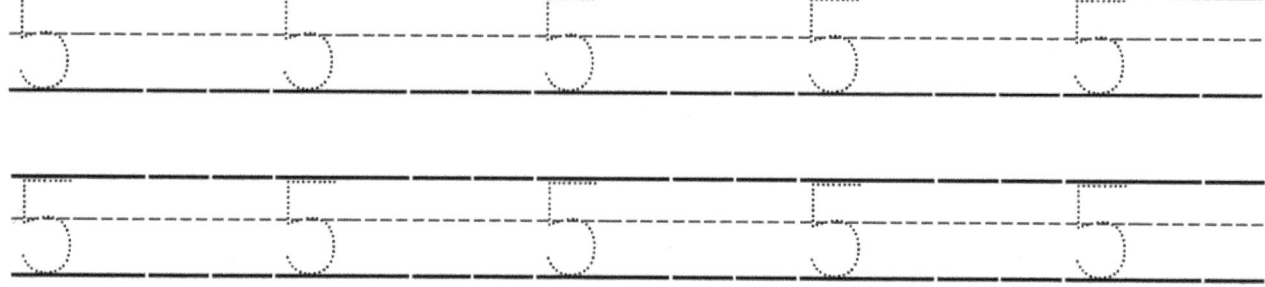

Circle Number 6s

Trace

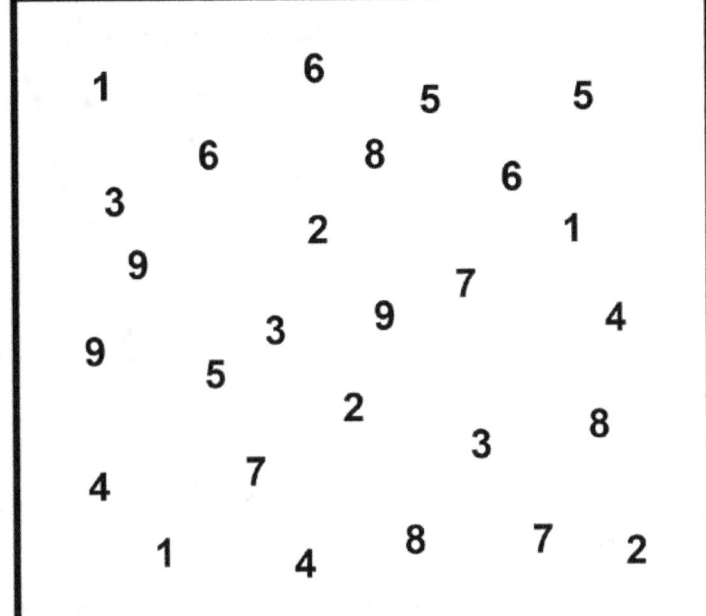
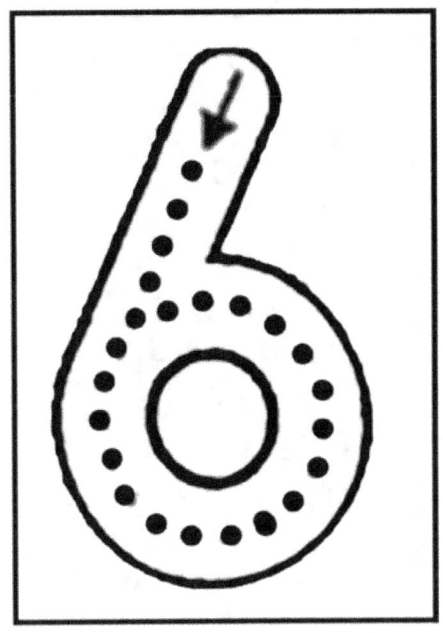

Color 6 Triangles

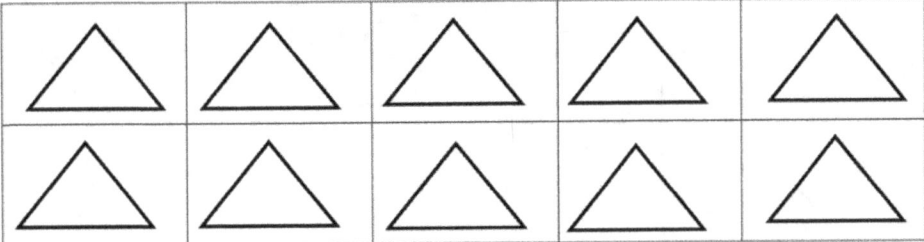

Tracing Number 6s

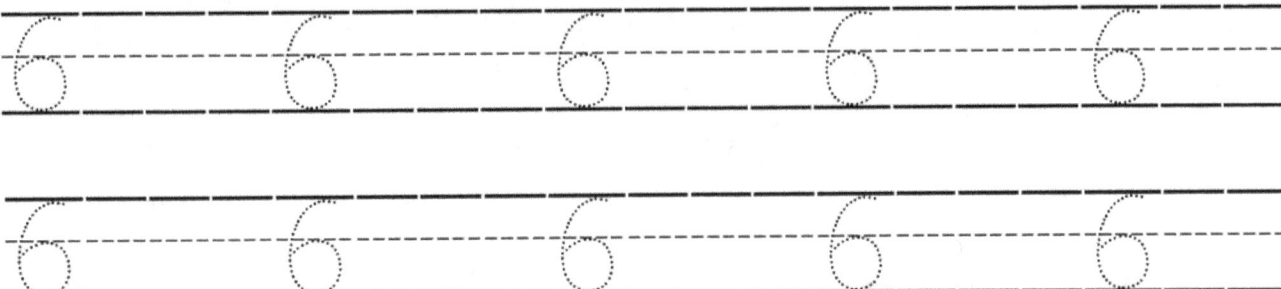

Circle Number 7s

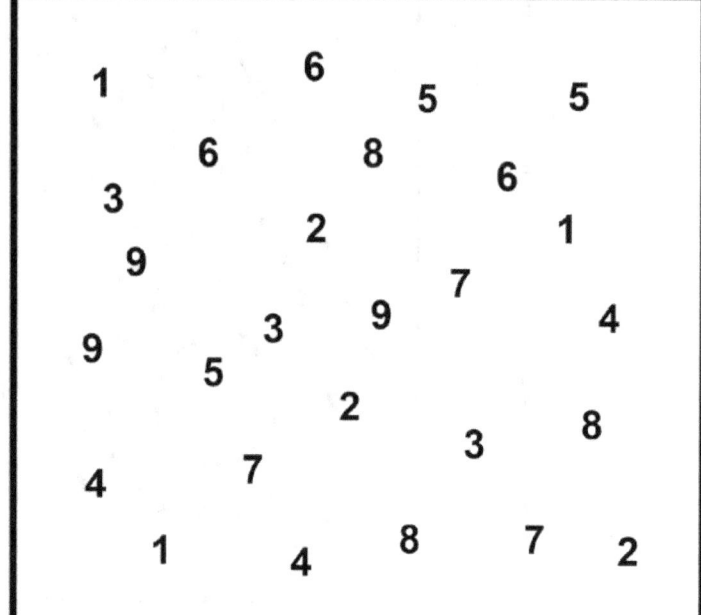

Trace

Color 7 Triangles

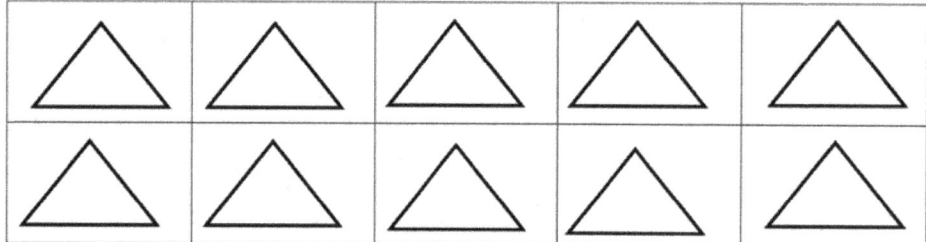

Tracing Number 6s

Circle Number 8s

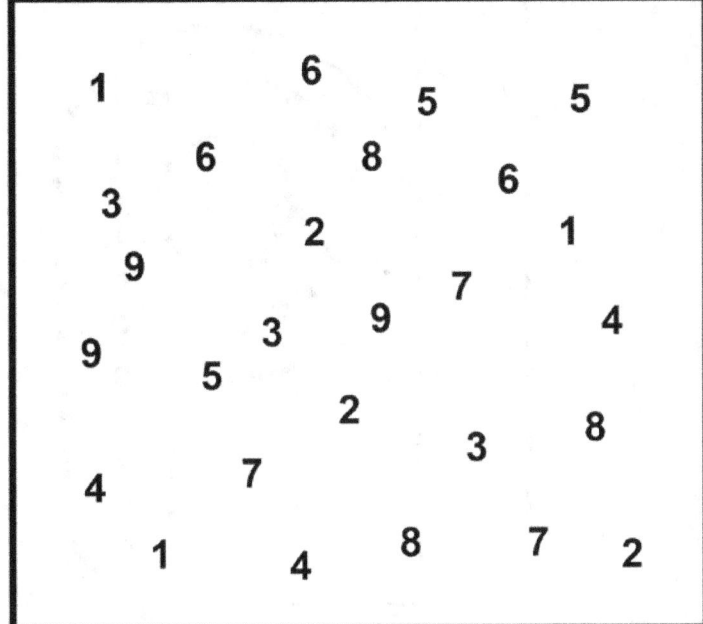

Trace

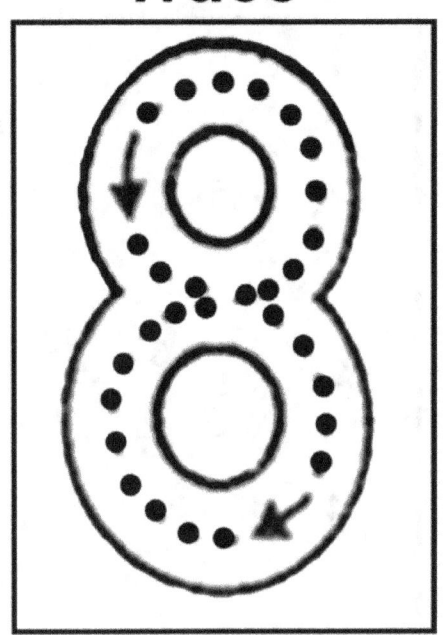

Color 8 Triangles

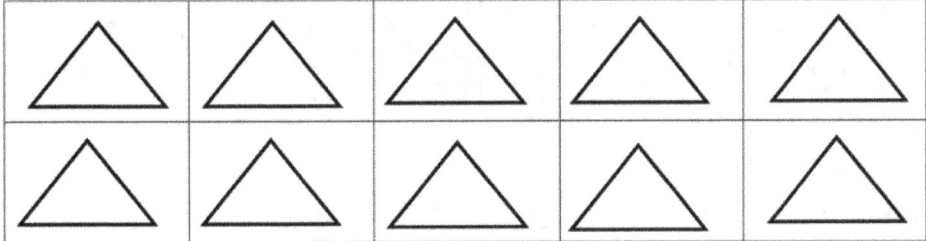

Tracing Number 8s

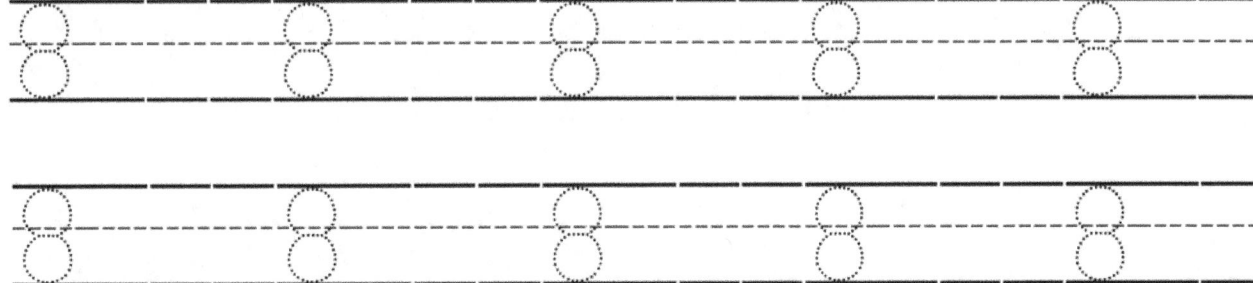

Circle Number 9s

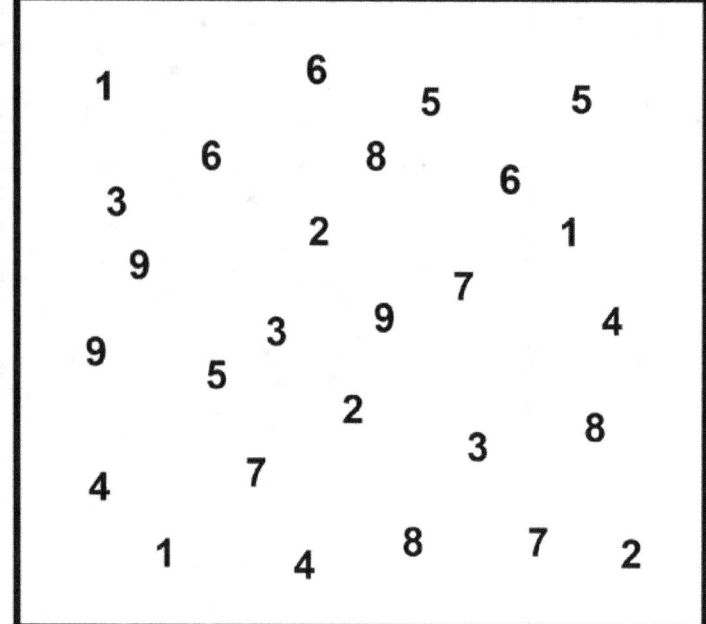

Trace

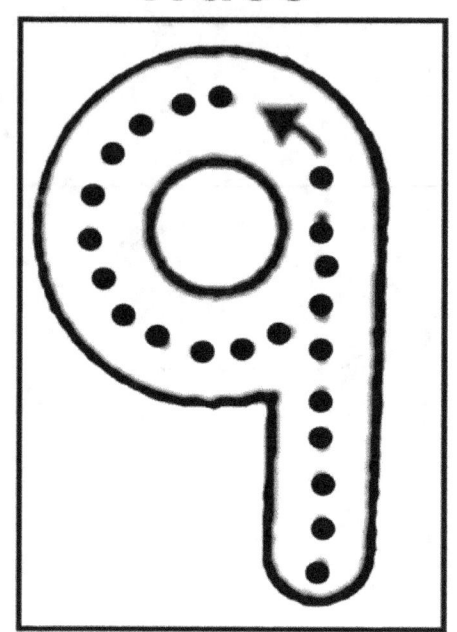

Color 9 Triangles

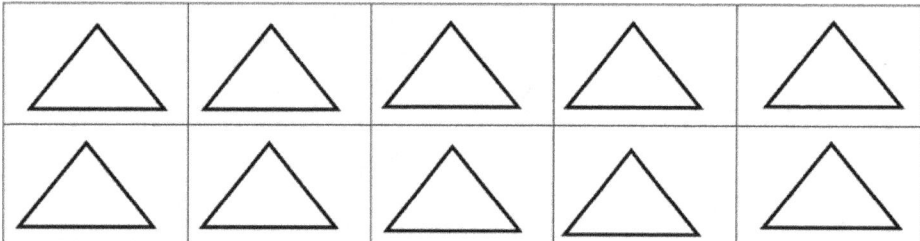

Tracing Number 9s

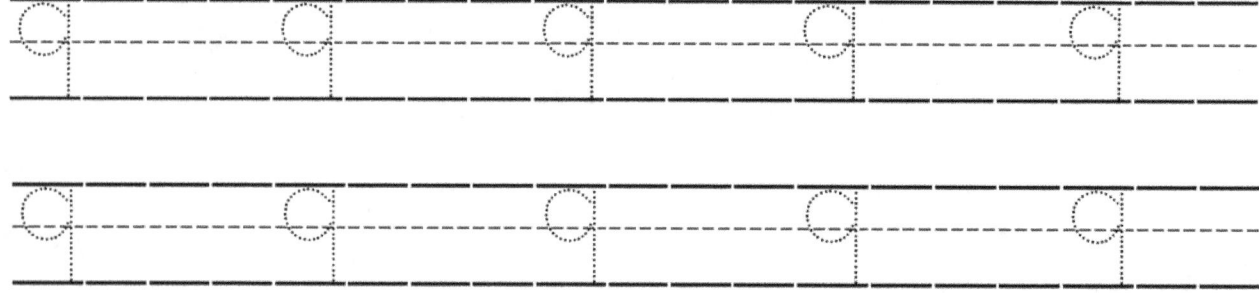

Count & Match

Count the objects & Match with the correct numbers

Count the objects & Match with the correct numbers

Count the objects & Match with the correct numbers

COUNT AND MATCH

count the balloon Match with the correct number

Count & Match

Count the objects & Match with the correct numbers

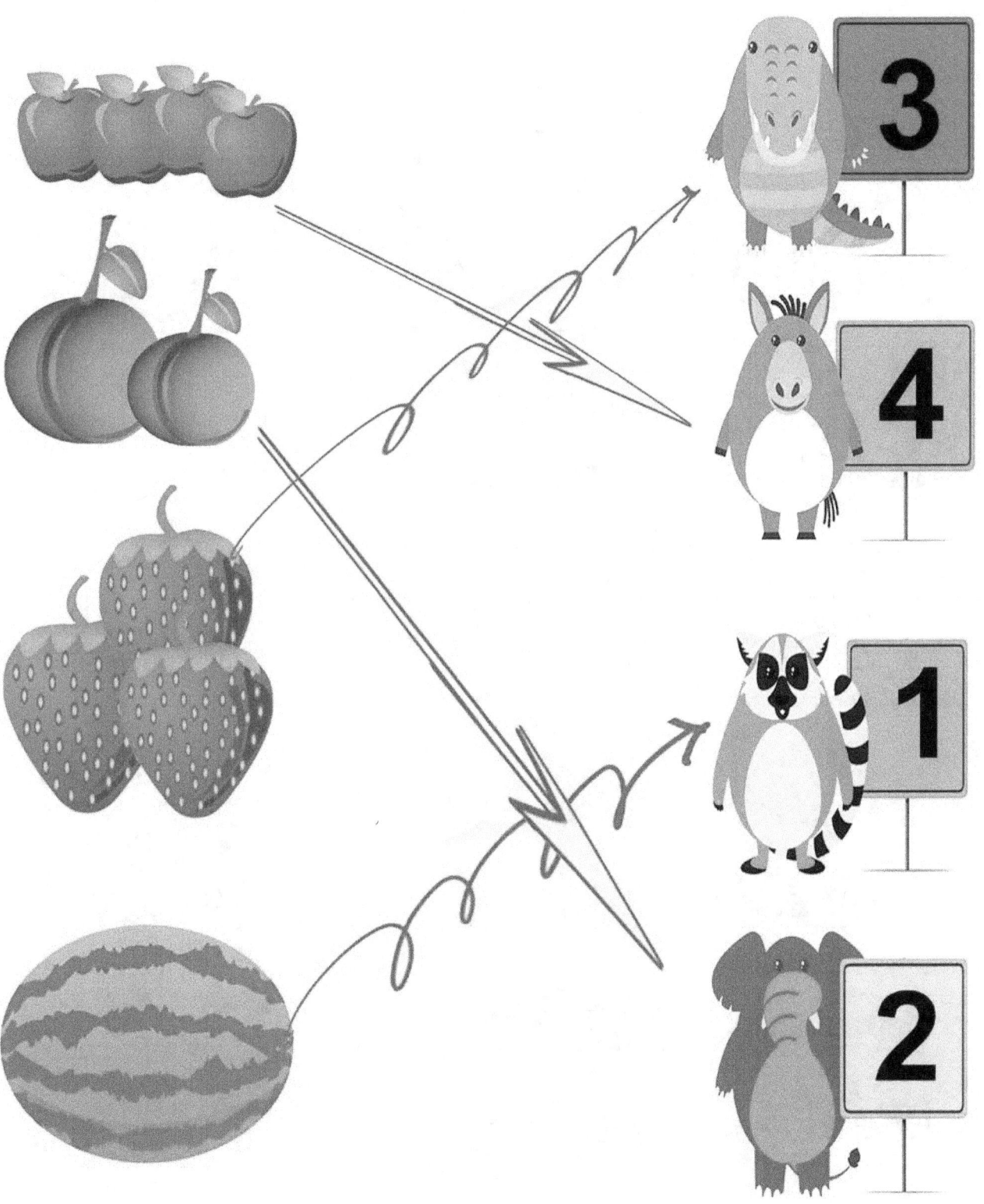

Count the objects & Match with the correct numbers

Count the objects & Match with the correct numbers

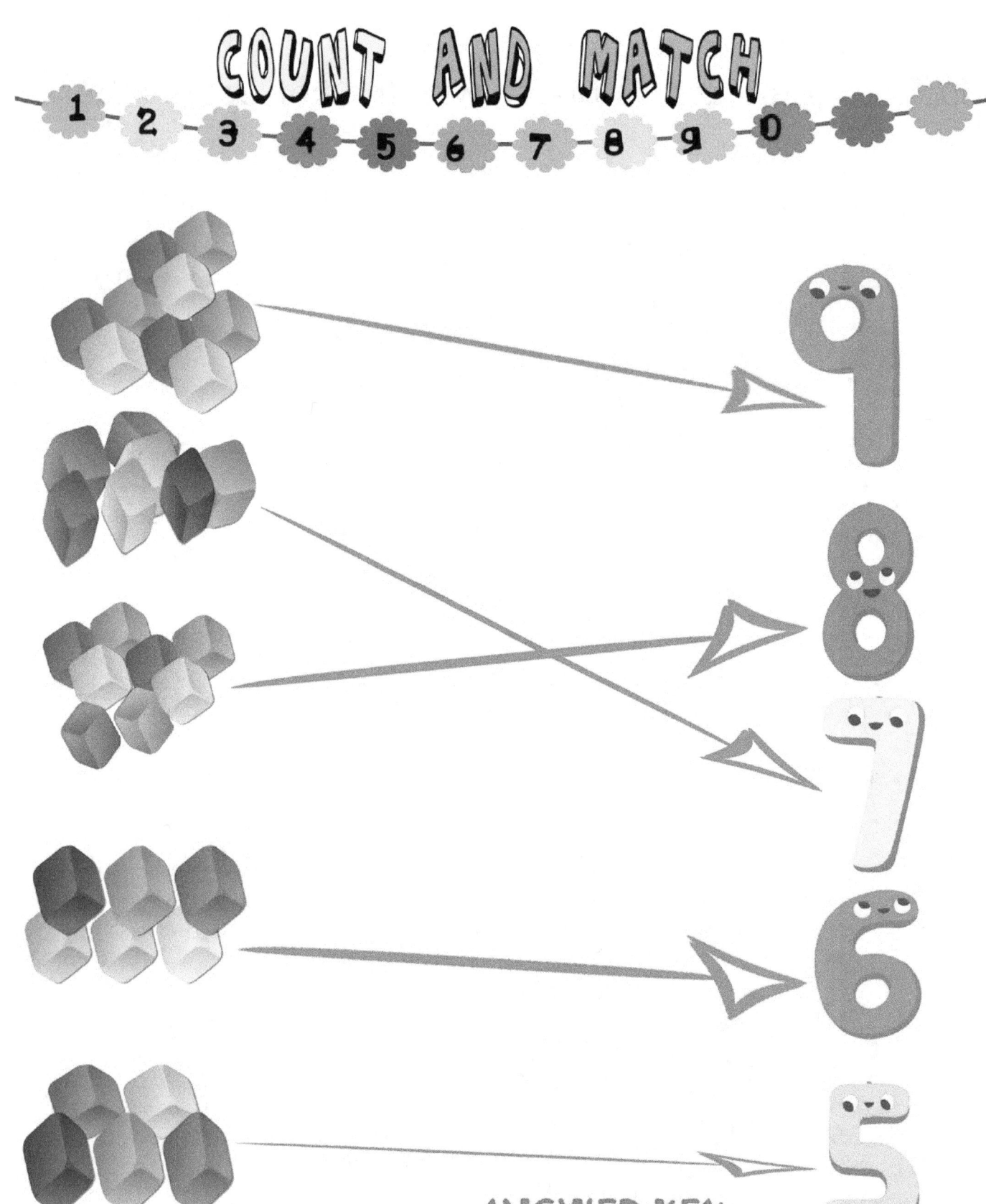

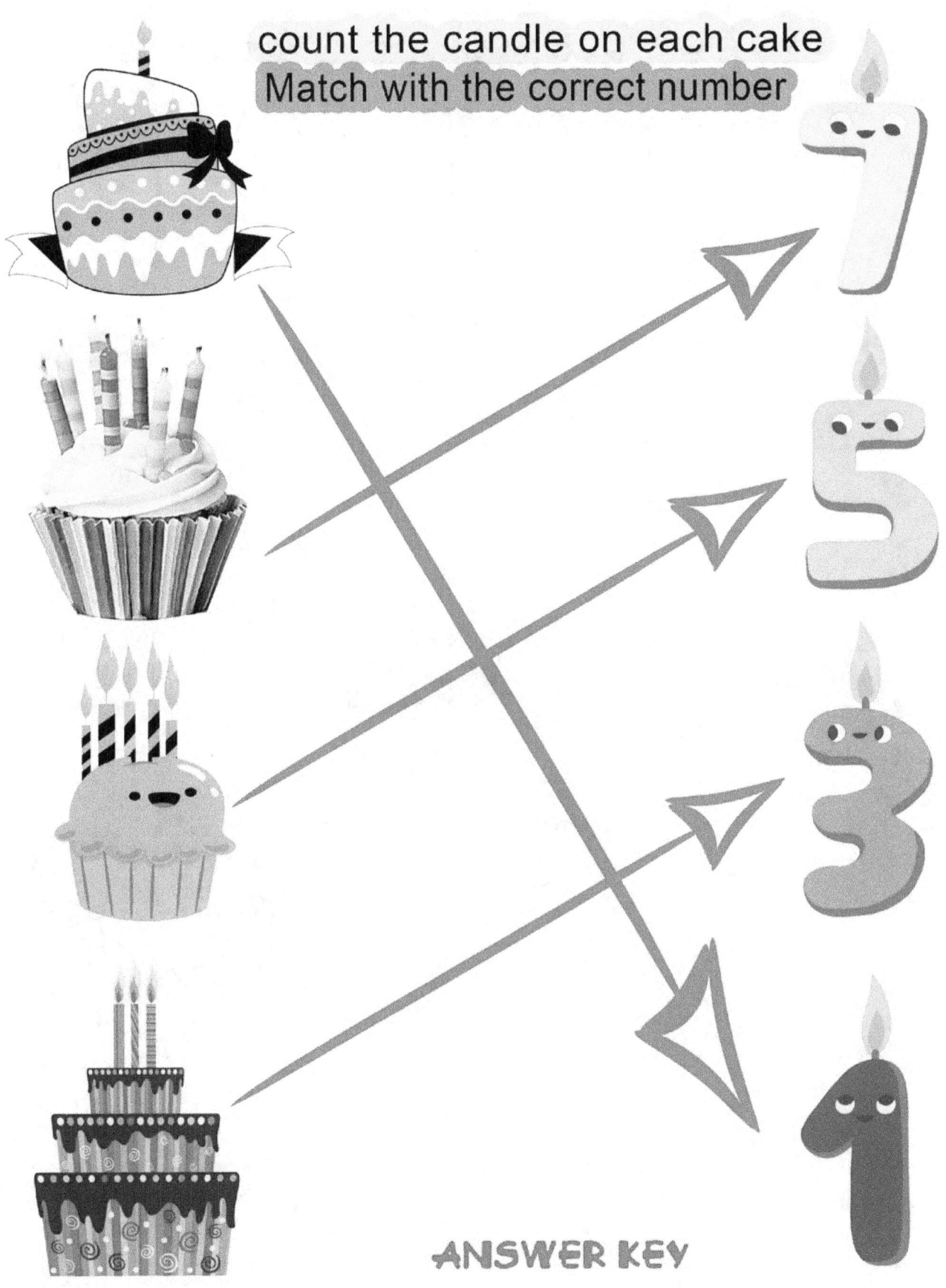

count the balloon   Match with the correct number

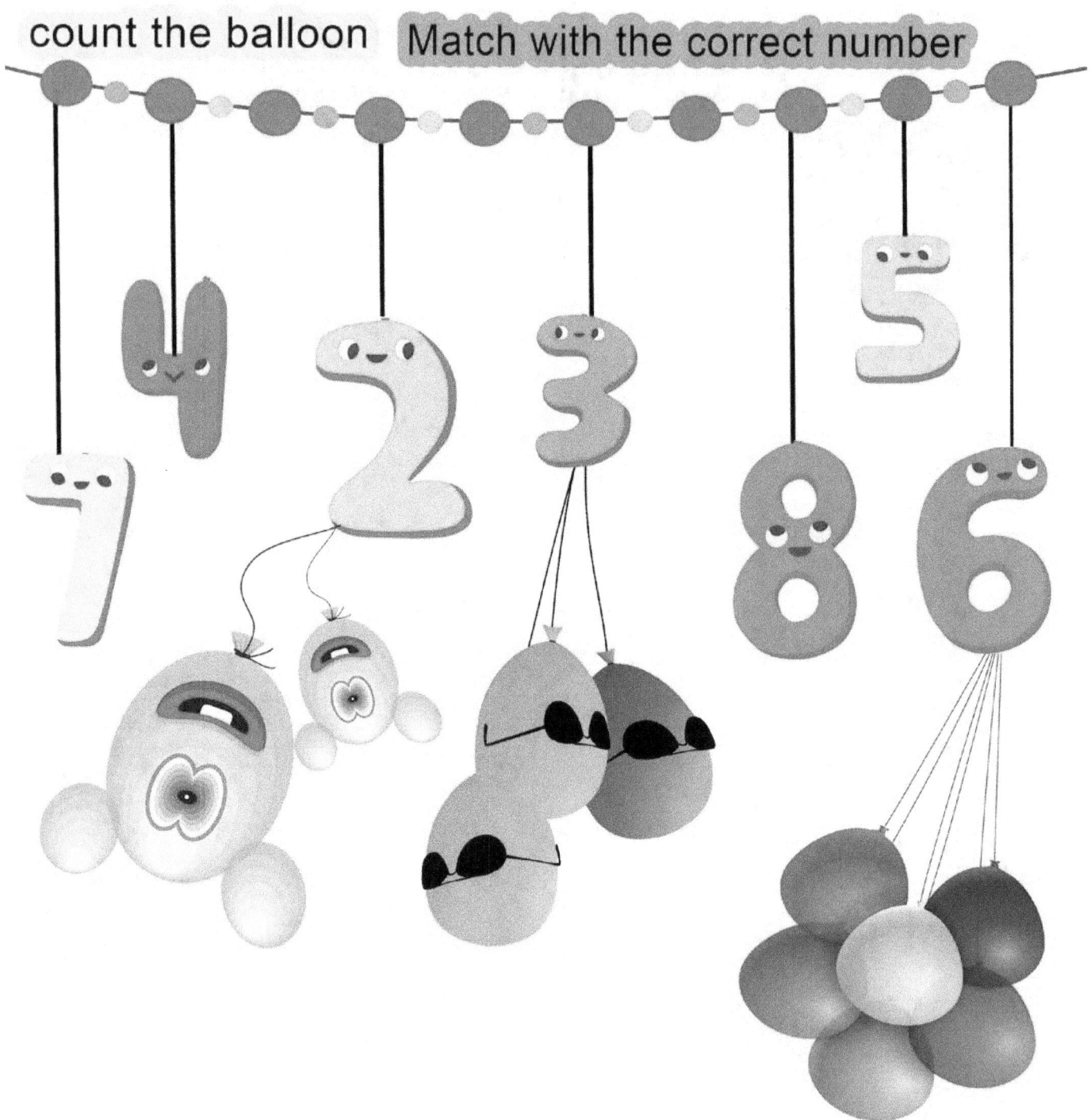

Answer Key

Your Score is

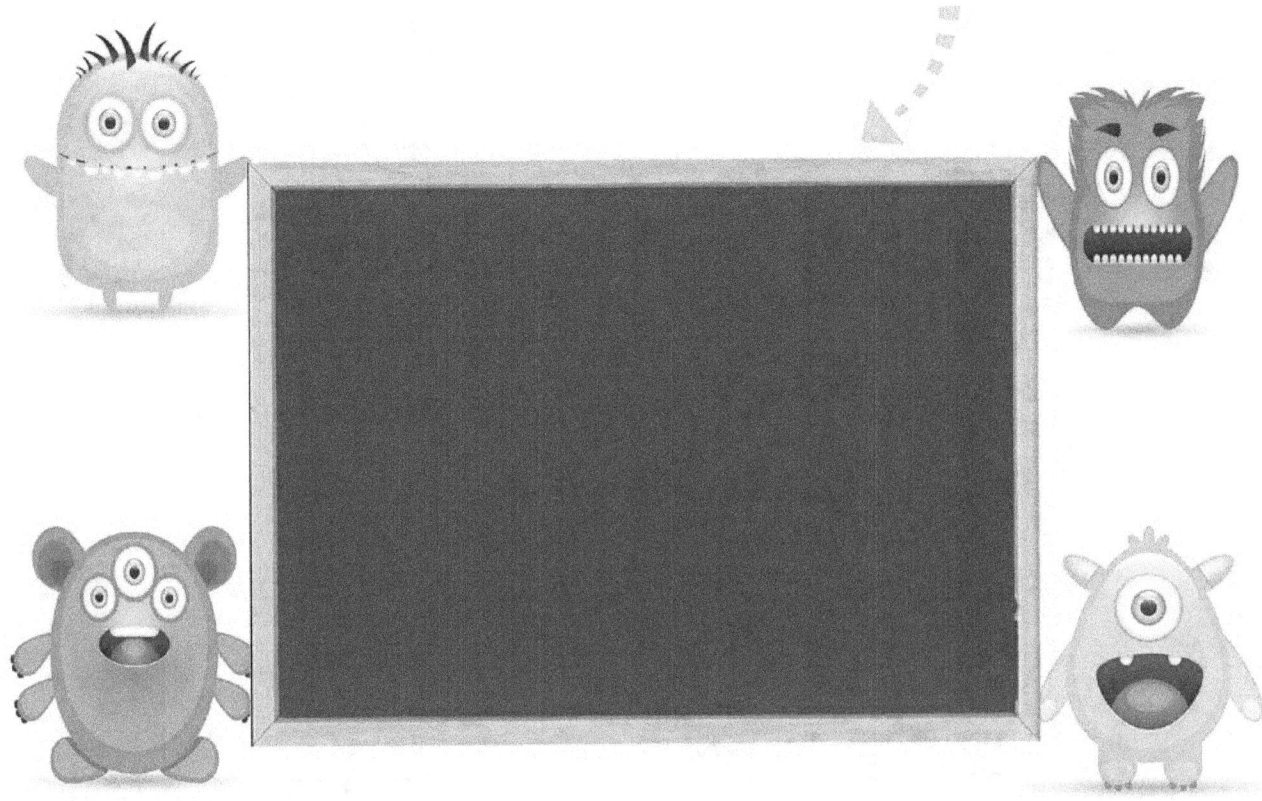

Note

Comparing Number

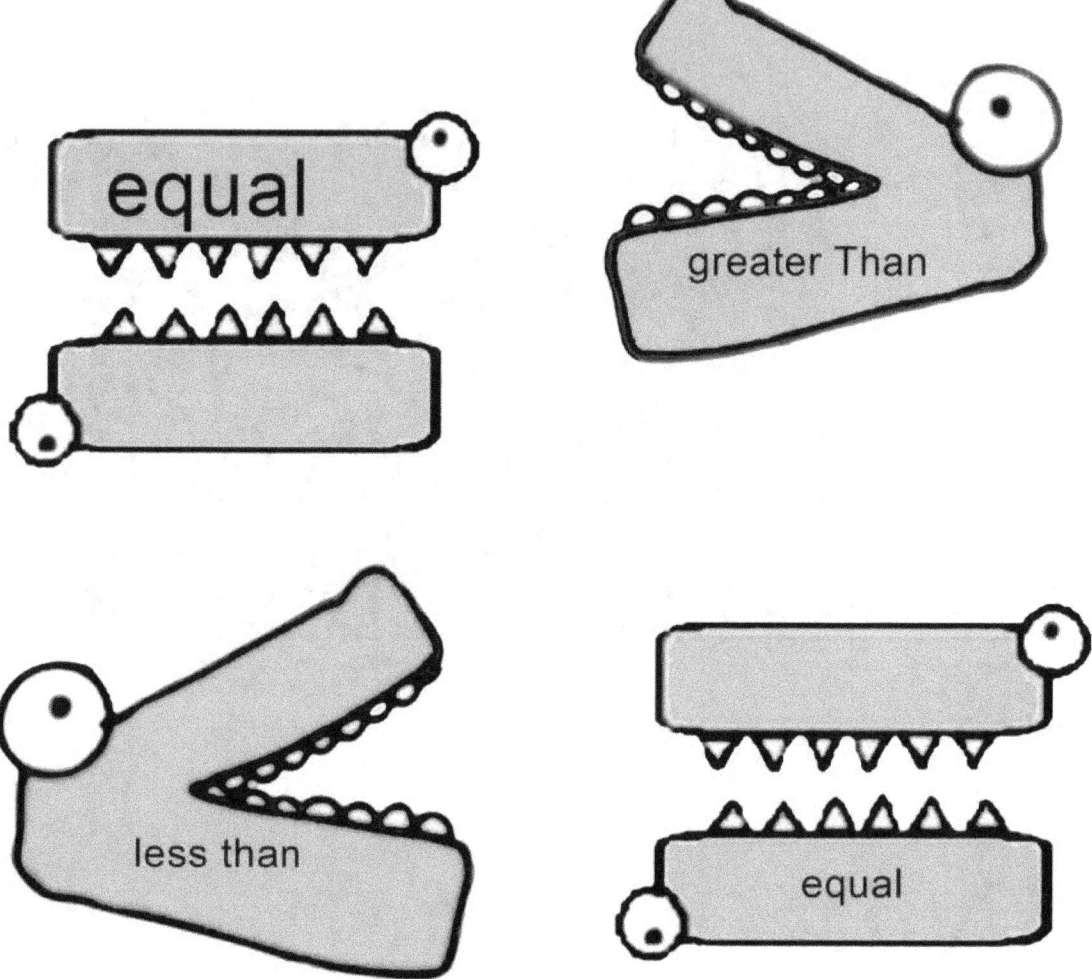

Greater than/ Less Than

Compare the groups in each row and write the correct sign in the middle

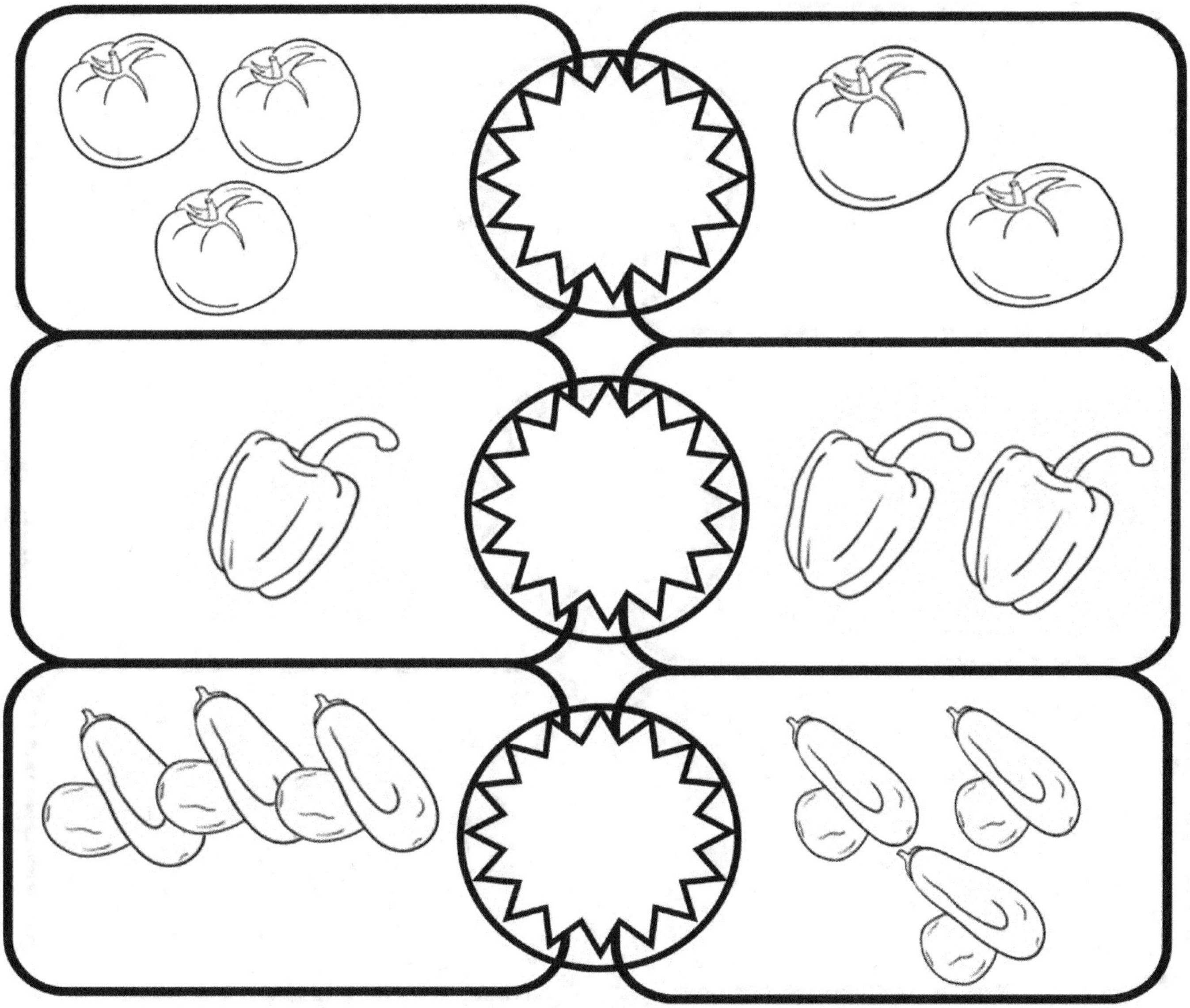

> = <

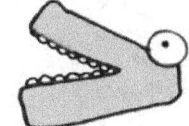 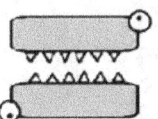 

Gearter than   Equal   Less Than

Greater than/ Less Than

Compare the groups in each row and write the correct sign in the middle

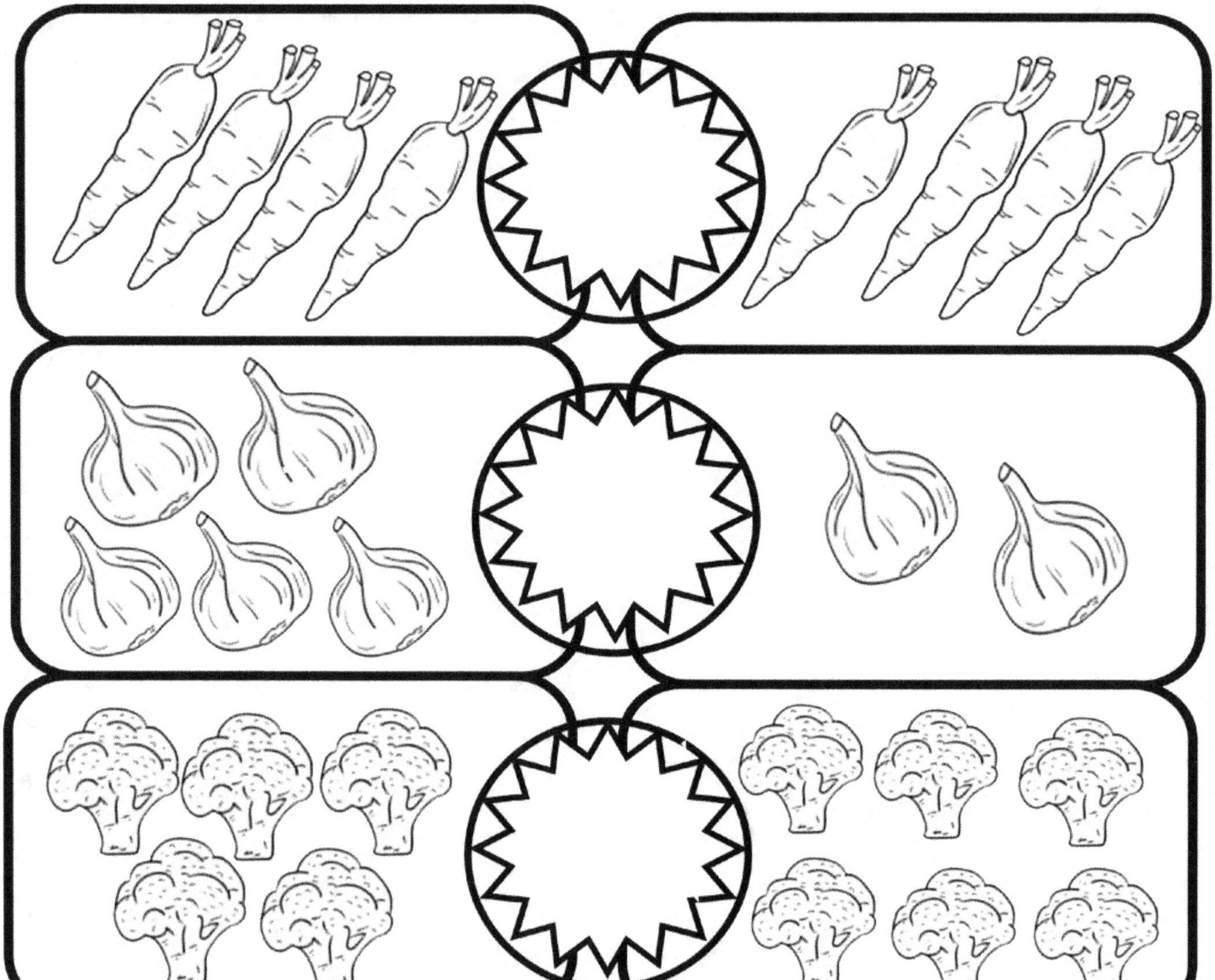

>   =   <

   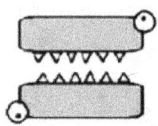   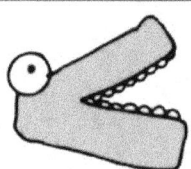

Gearter than   Equal   Less Than

 # Greater than/ Less Than

Compare the groups in each row and write the correct sign in the middle

> = <

Gearter than   Equal   Less Than

 # Greater than/ Less Than

Compare the groups in each row and write the correct sign in the middle

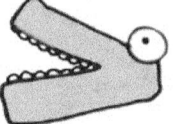   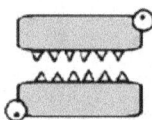   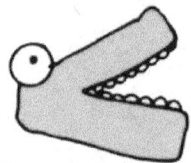

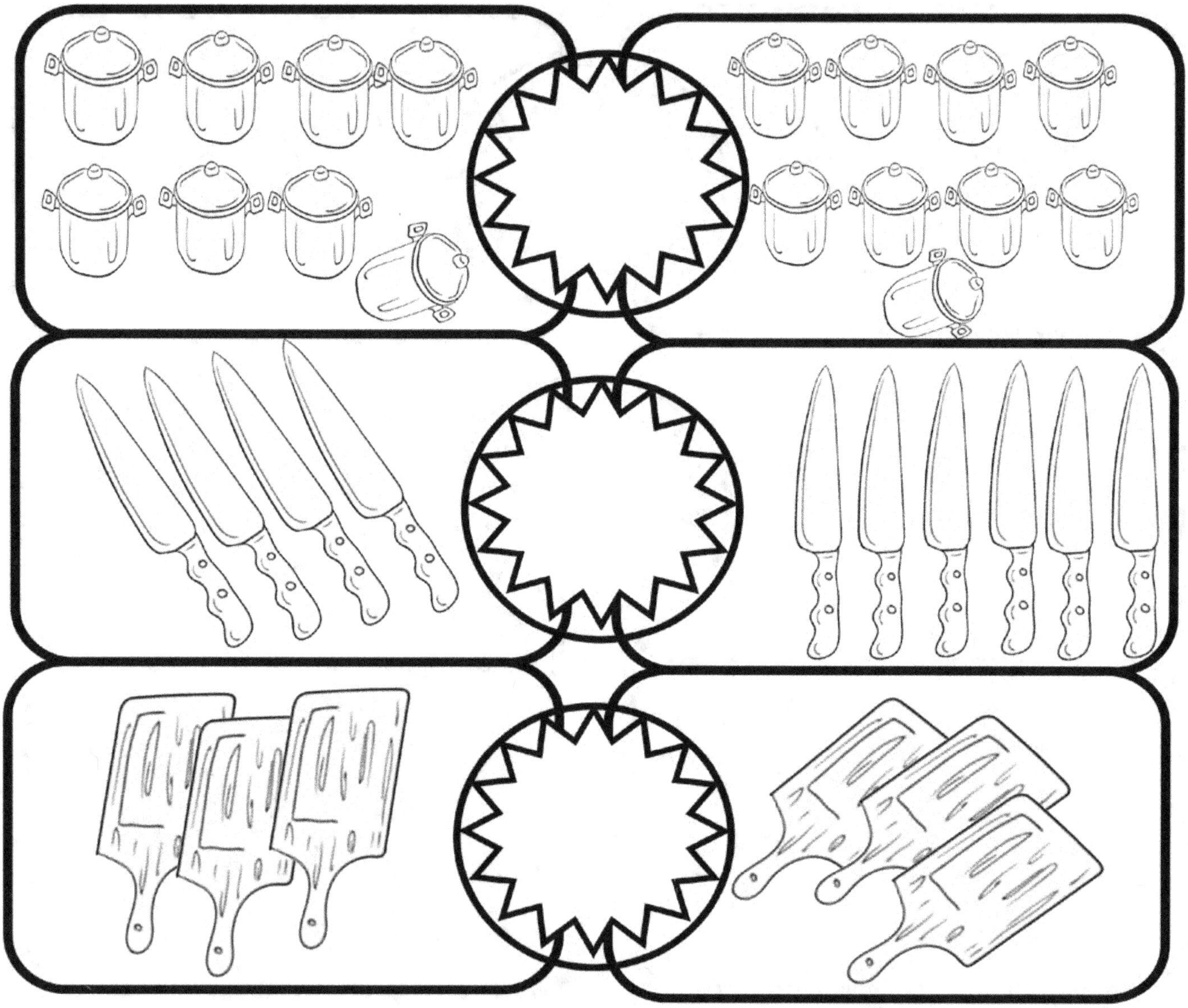

Gearter than   Equal   Less Than

Greater than/ Less Than

Compare the groups in each row and write the correct sign in the middle

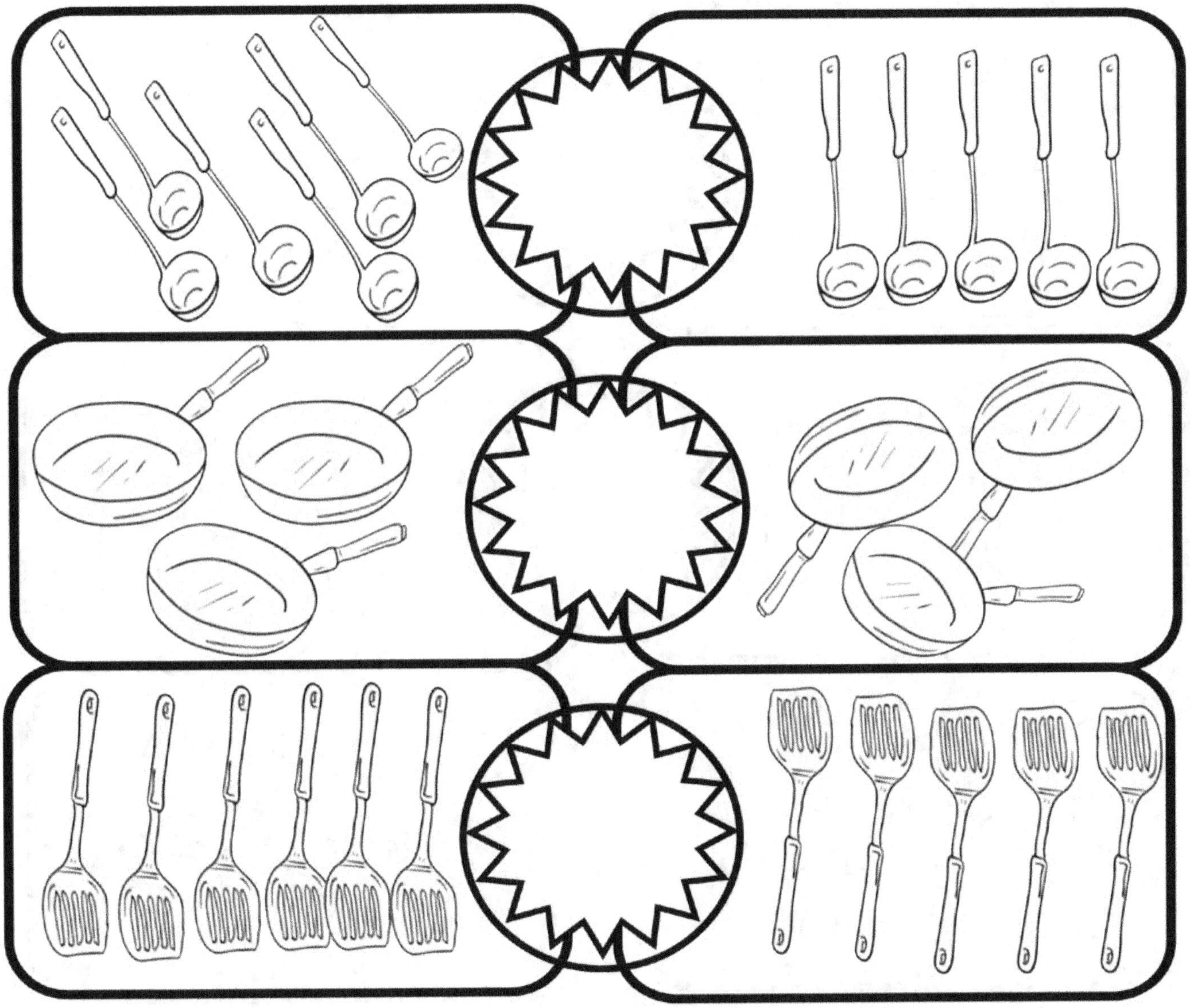

>   =   <

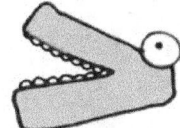
Gearter than

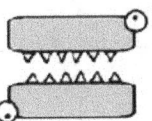
Equal

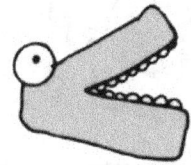
Less Than

Greater than/ Less Than

Compare the groups in each row and write the correct sign in the middle

 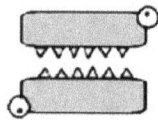 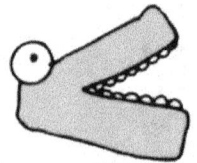

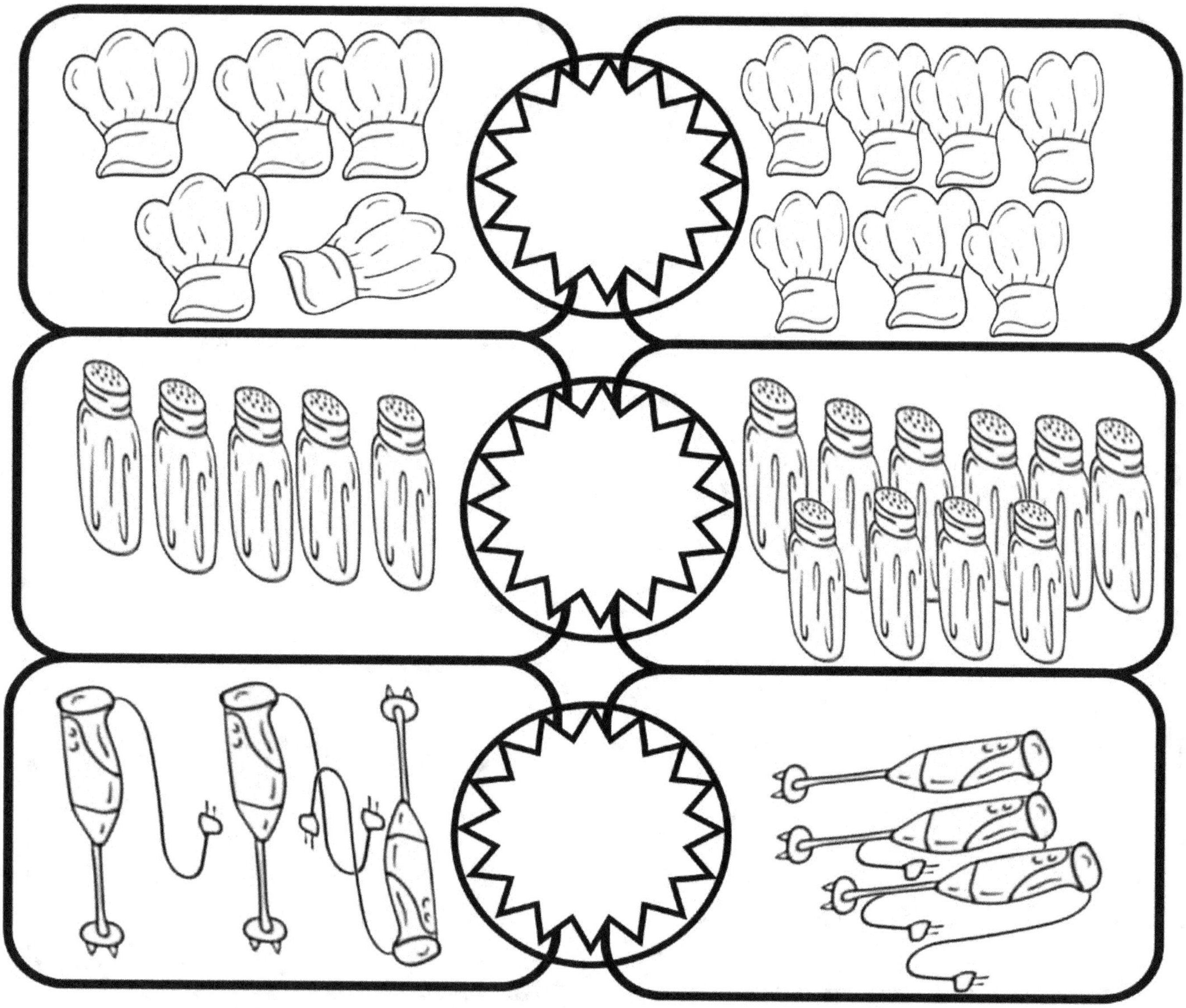

Gearter than   Equal   Less Than

Greater than/ Less Than

Compare the groups in each row and write the correct sign in the middle

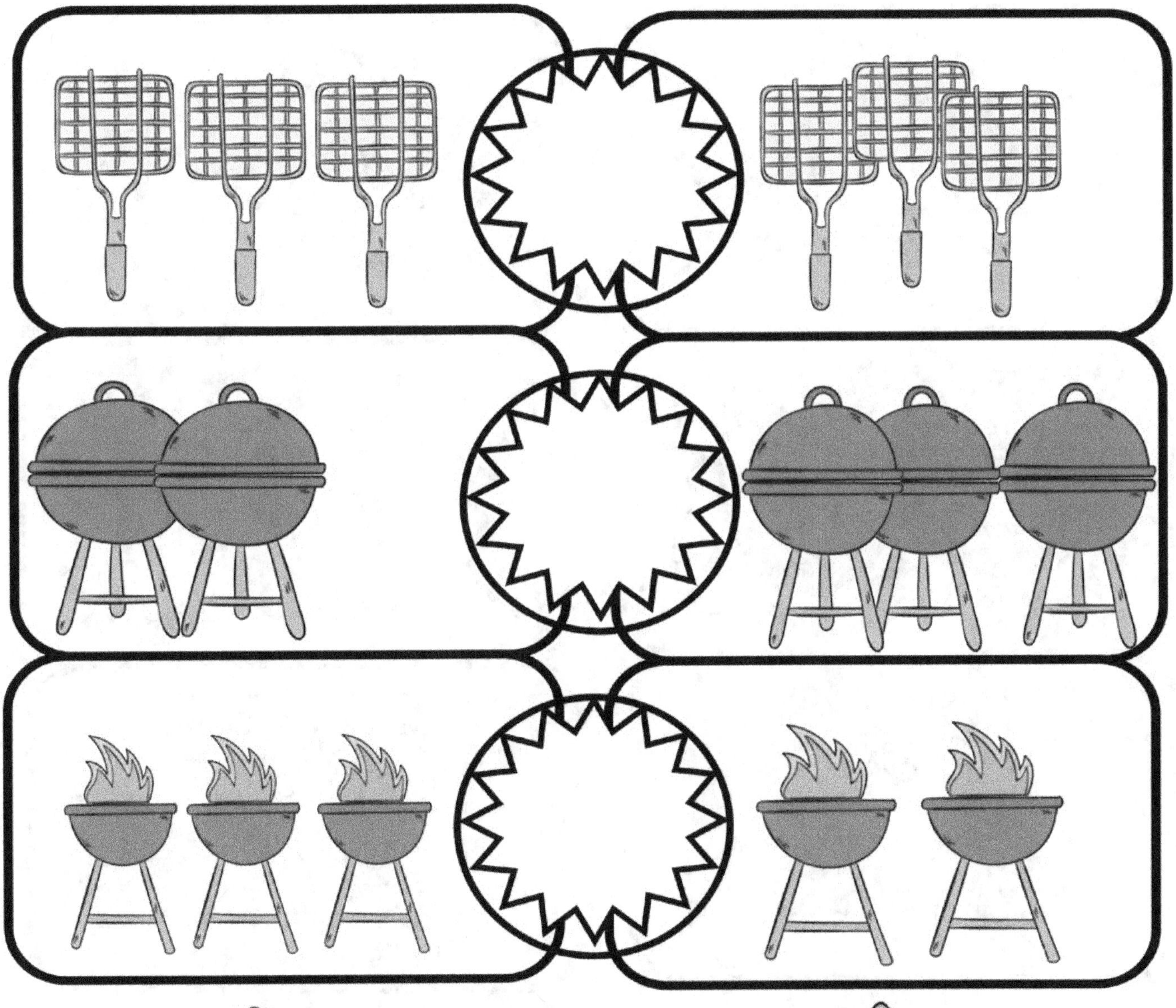

> = <

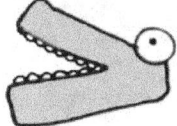   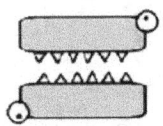   

Gearter than   Equal   Less Than

Greater than/ Less Than

Compare the groups in each row and write the correct sign in the middle

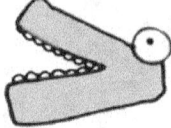 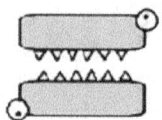 

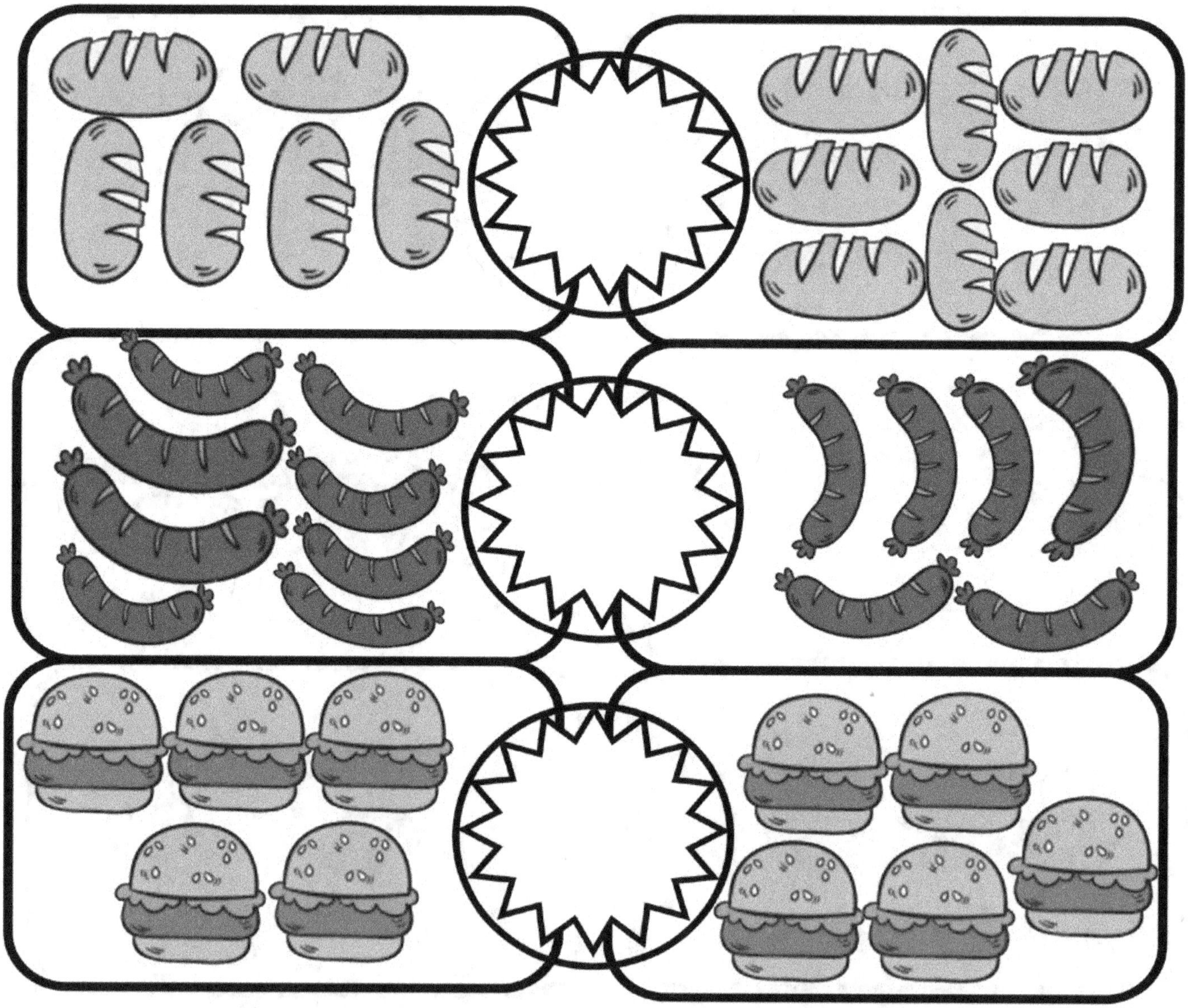

Gearter than   Equal   Less Than

Greater than/ Less Than

Compare the groups in each row and write the correct sign in the middle

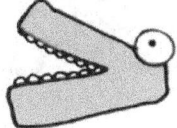 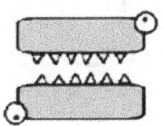 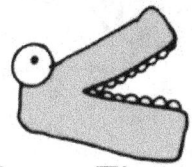

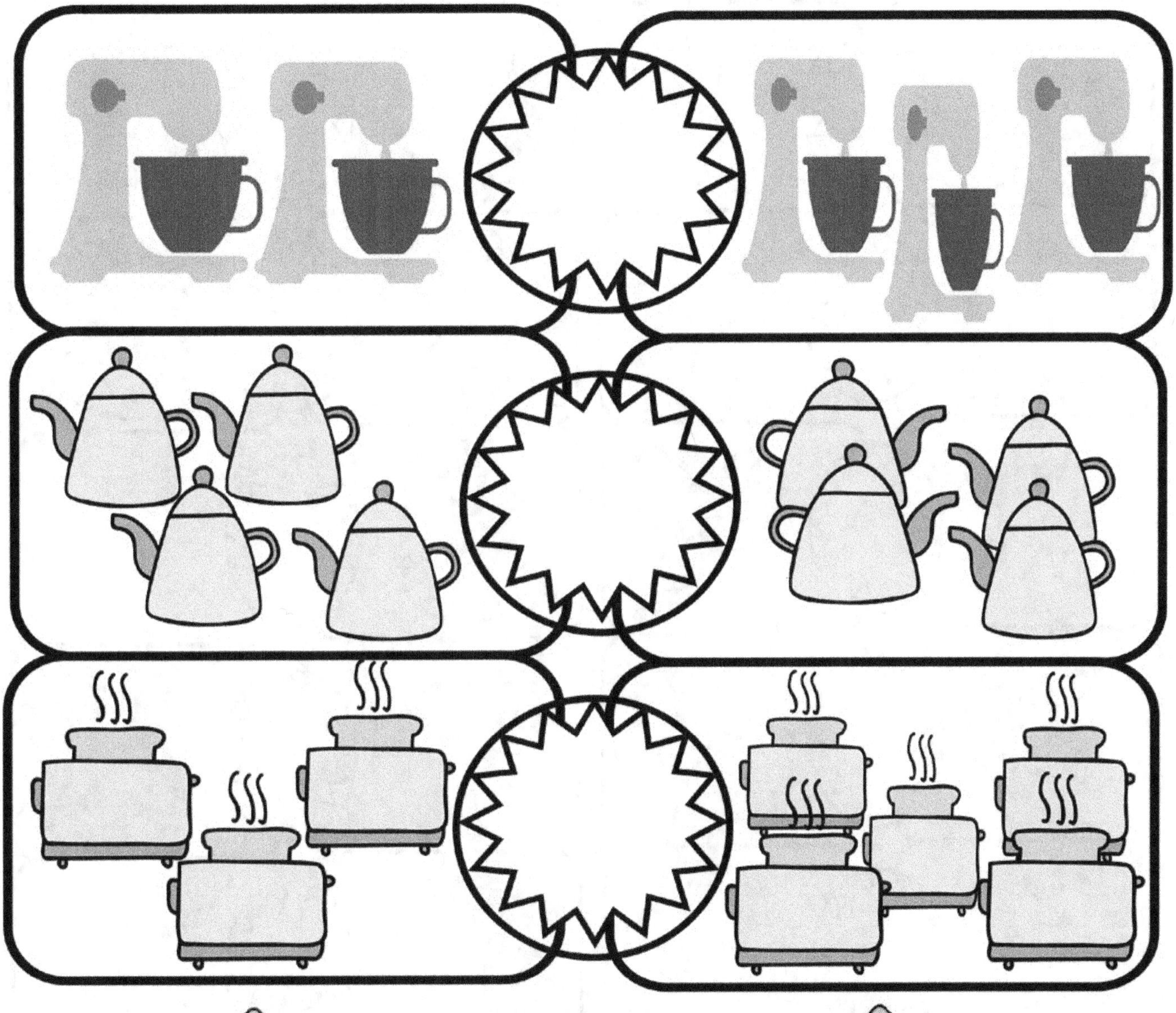

Gearter than   Equal   Less Than

Greater than/ Less Than

Comparing the number
and write the correct sign in the middle

> = <

Greater than/ Less Than

Comparing the number and write the correct sign in the middle

> = <

Greater than/ Less Than

Comparing the number and write the correct sign in the middle

> = <

Comparing Number

Answer

Greater than/ Less Than

Compare the groups in each row and write the correct sign in the middle

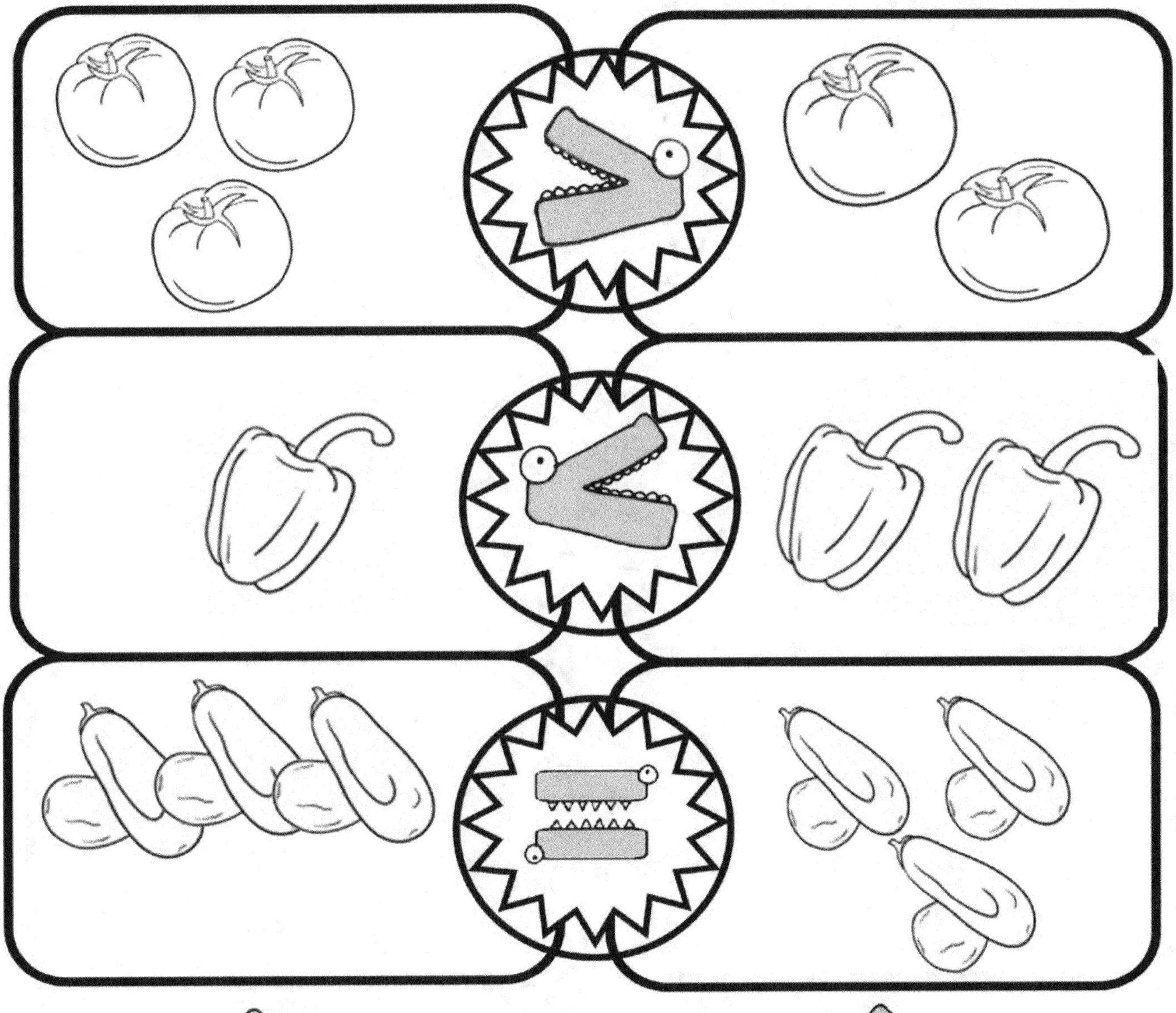

Gearter than   Equal   Less Than

Greater than/ Less Than

Compare the groups in each row and write the correct sign in the middle

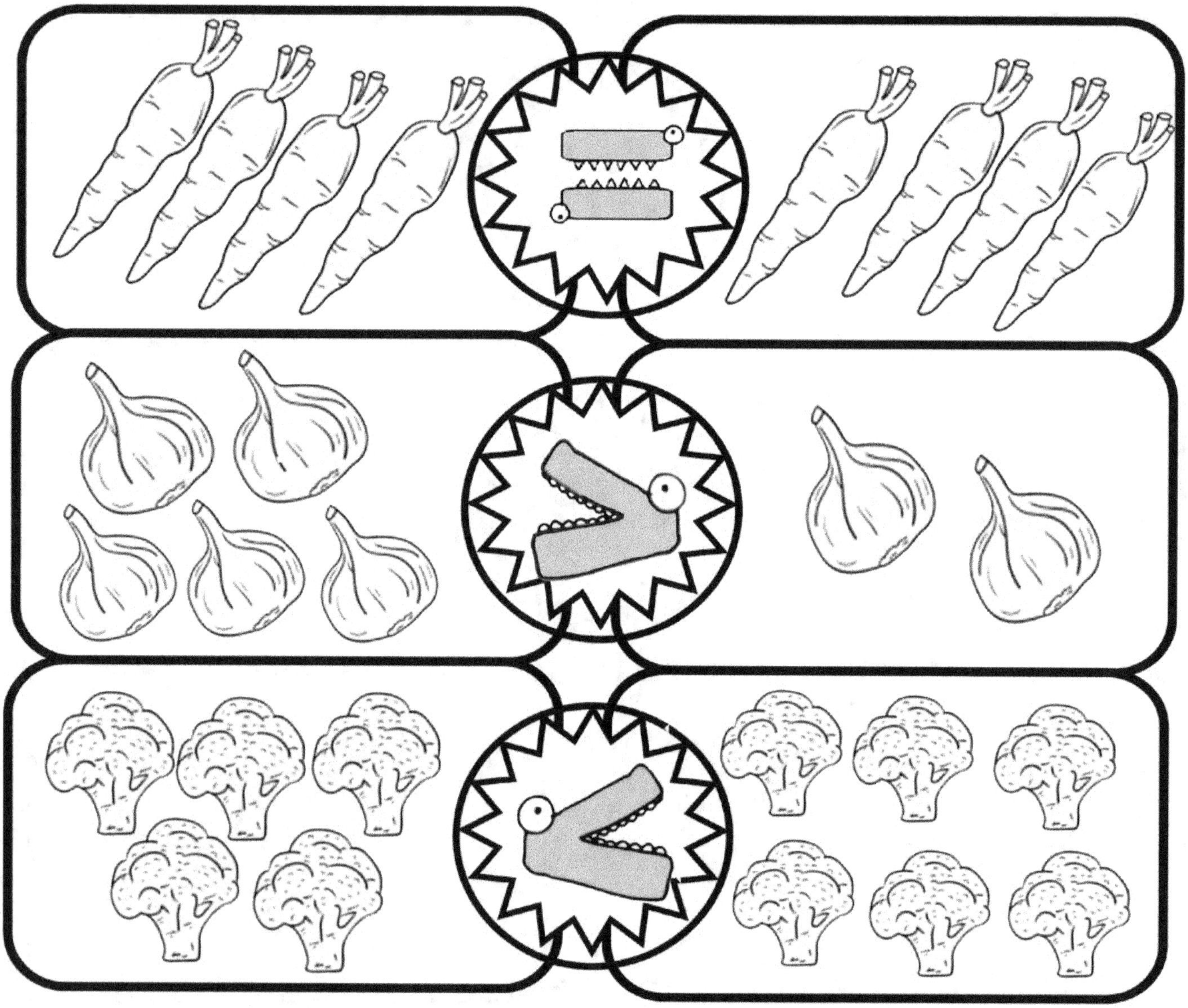

	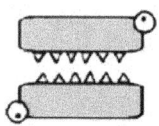	
Gearter than	Equal	Less Than

 # Greater than/ Less Than

Compare the groups in each row and write the correct sign in the middle

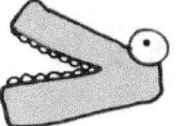 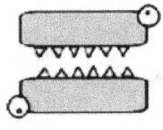 

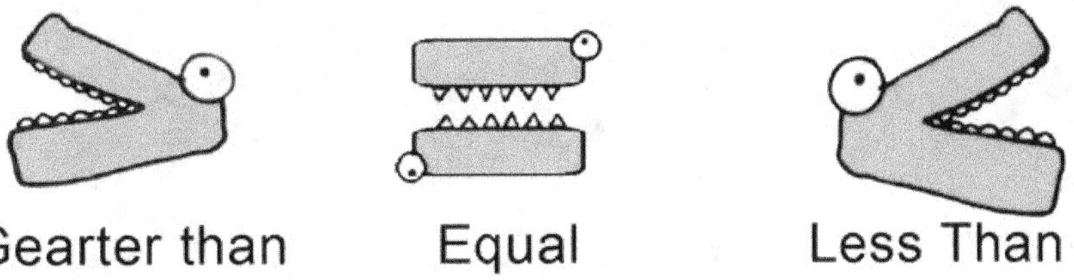

Gearter than   Equal   Less Than

# Greater than/ Less Than

Compare the groups in each row and write the correct sign in the middle

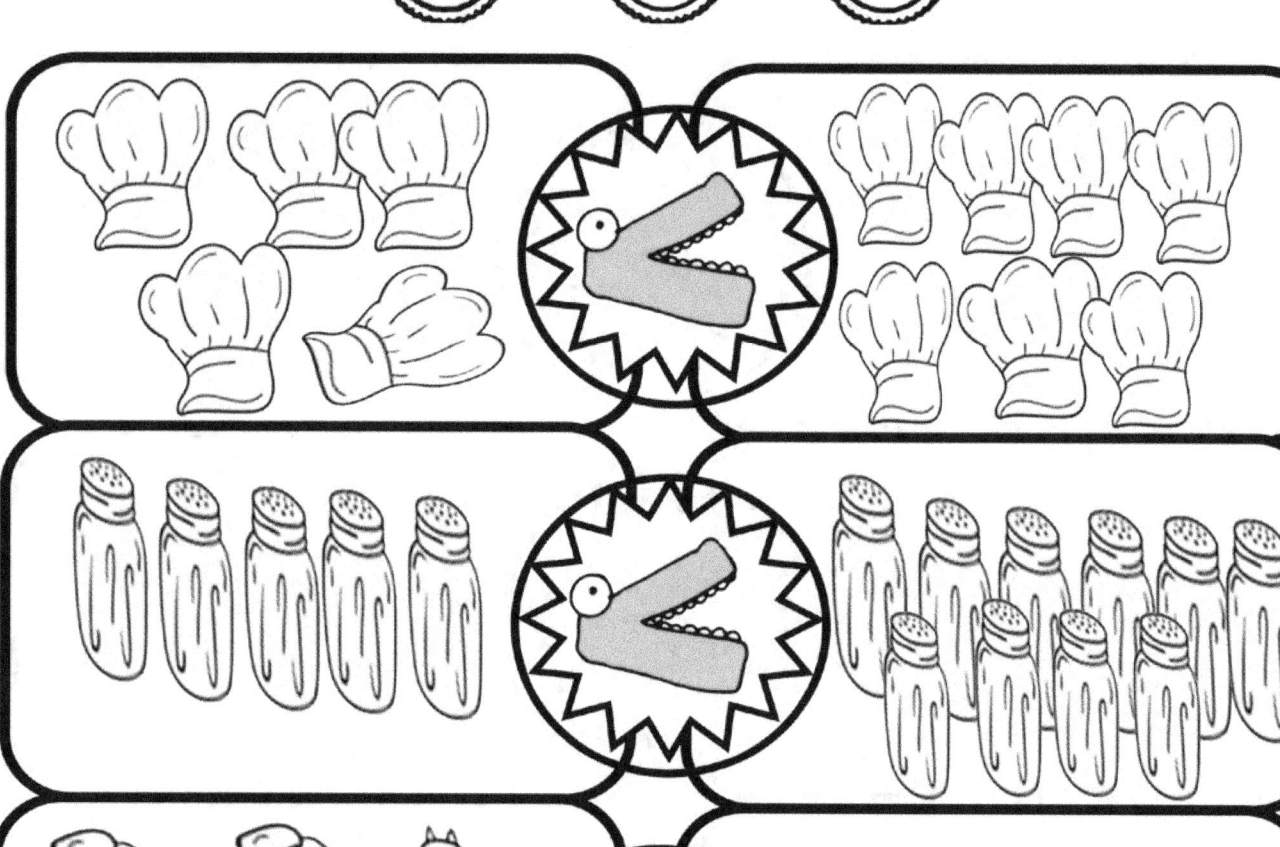

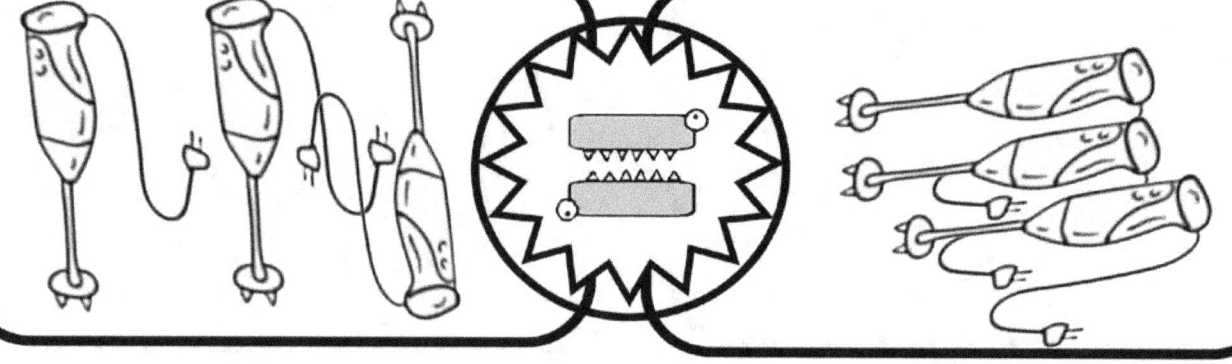

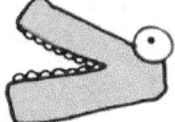 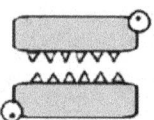 

Gearter than   Equal   Less Than

Greater than/ Less Than

Compare the groups in each row and write the correct sign in the middle

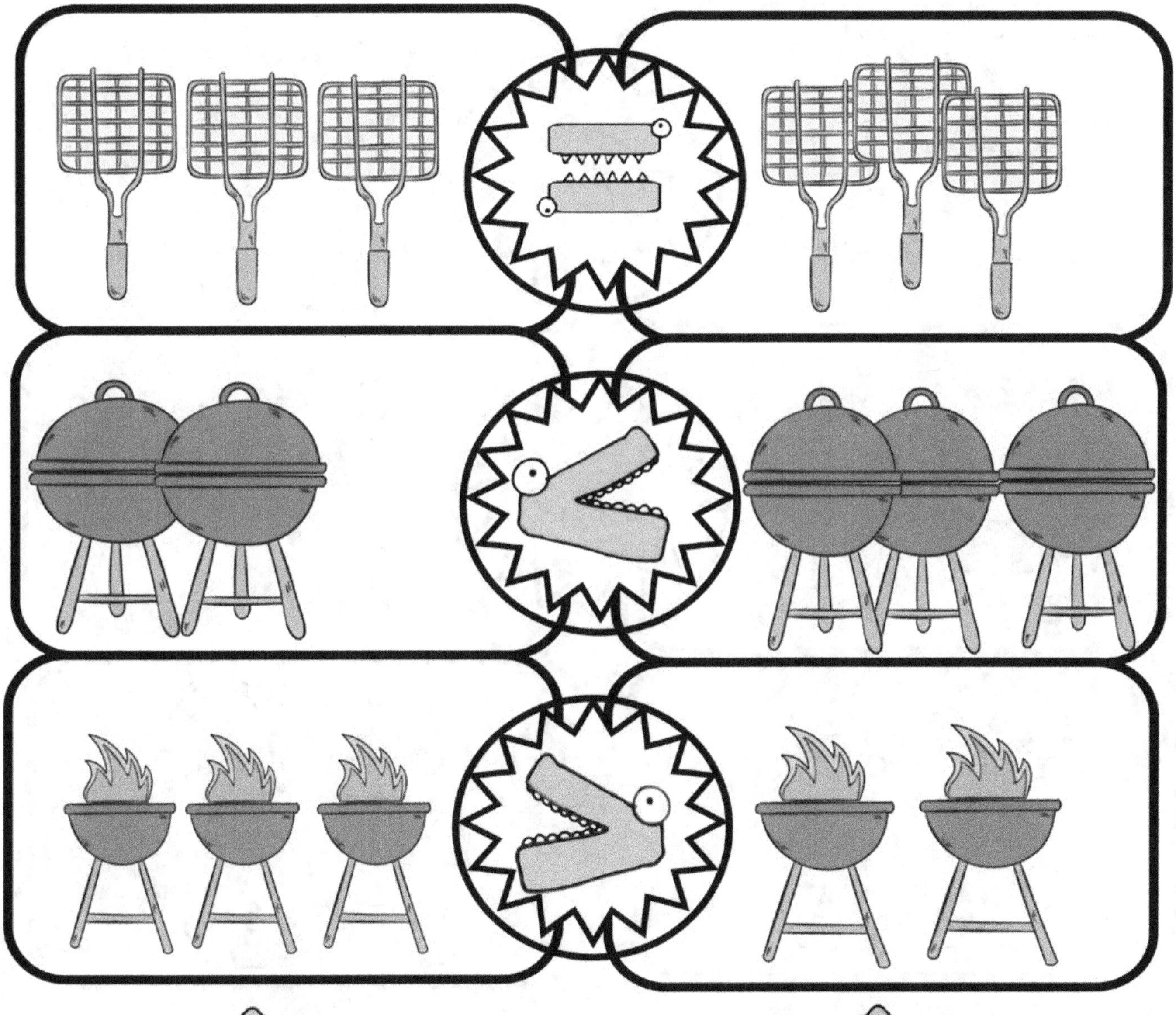

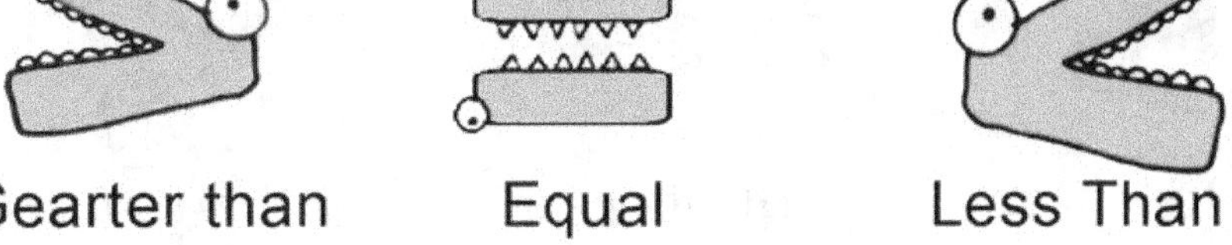

> = <

Gearter than   Equal   Less Than

Greater than/ Less Than

Compare the groups in each row and write the correct sign in the middle

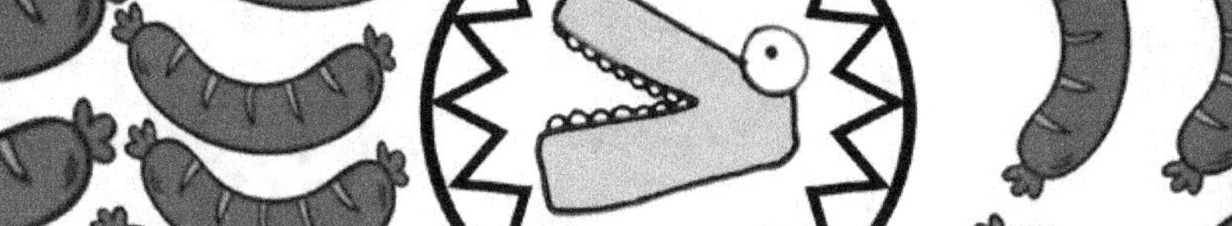

Gearter than   Equal   Less Than

# Greater than/ Less Than

Compare the groups in each row and write the correct sign in the middle

> = <

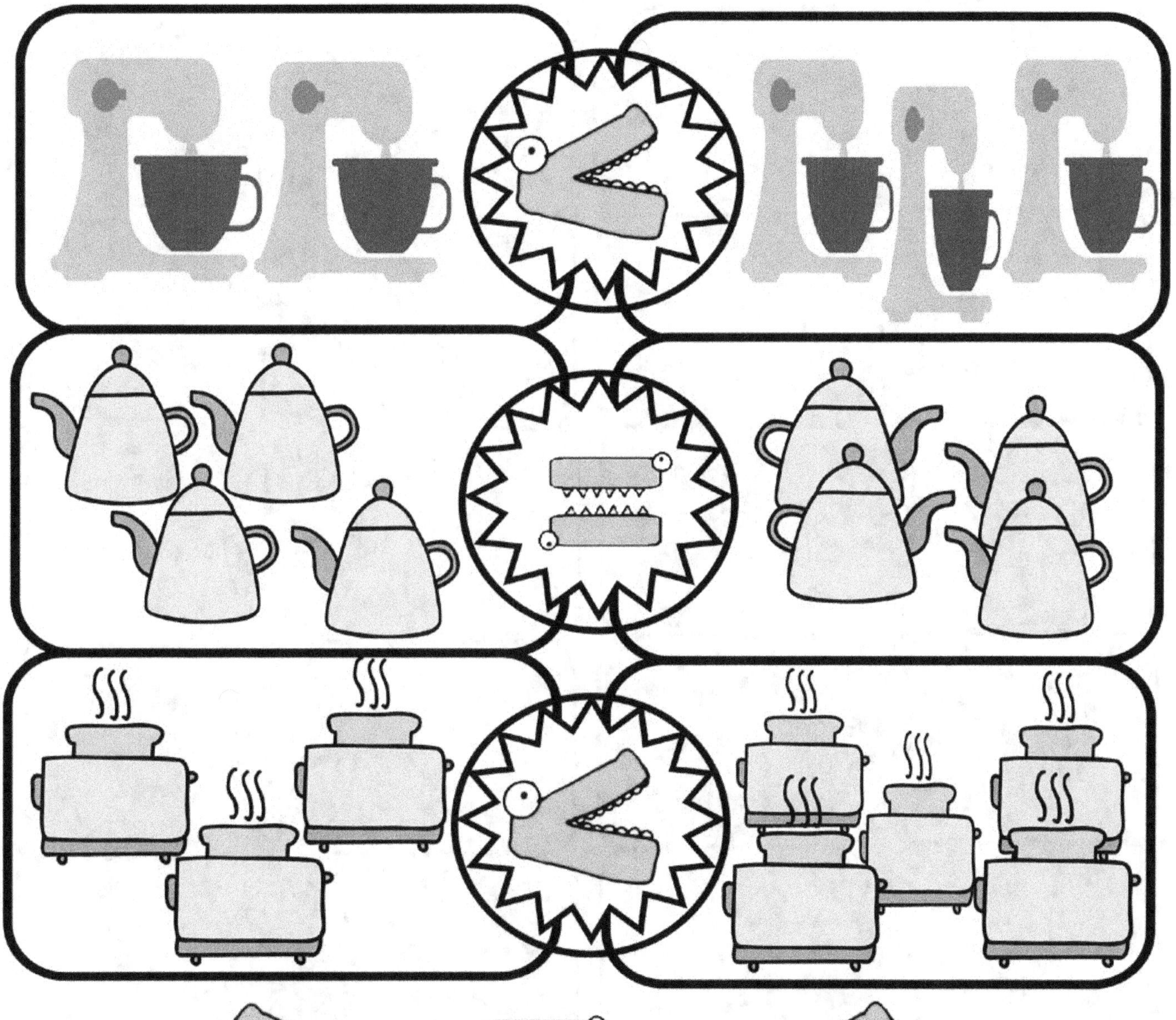

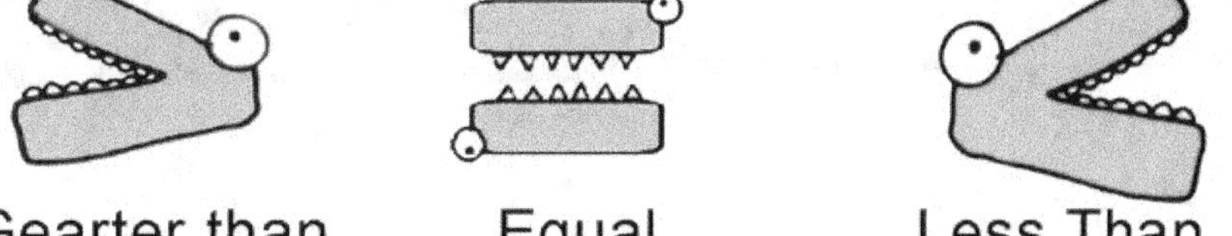

 Gearter than   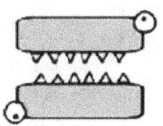 Equal    Less Than

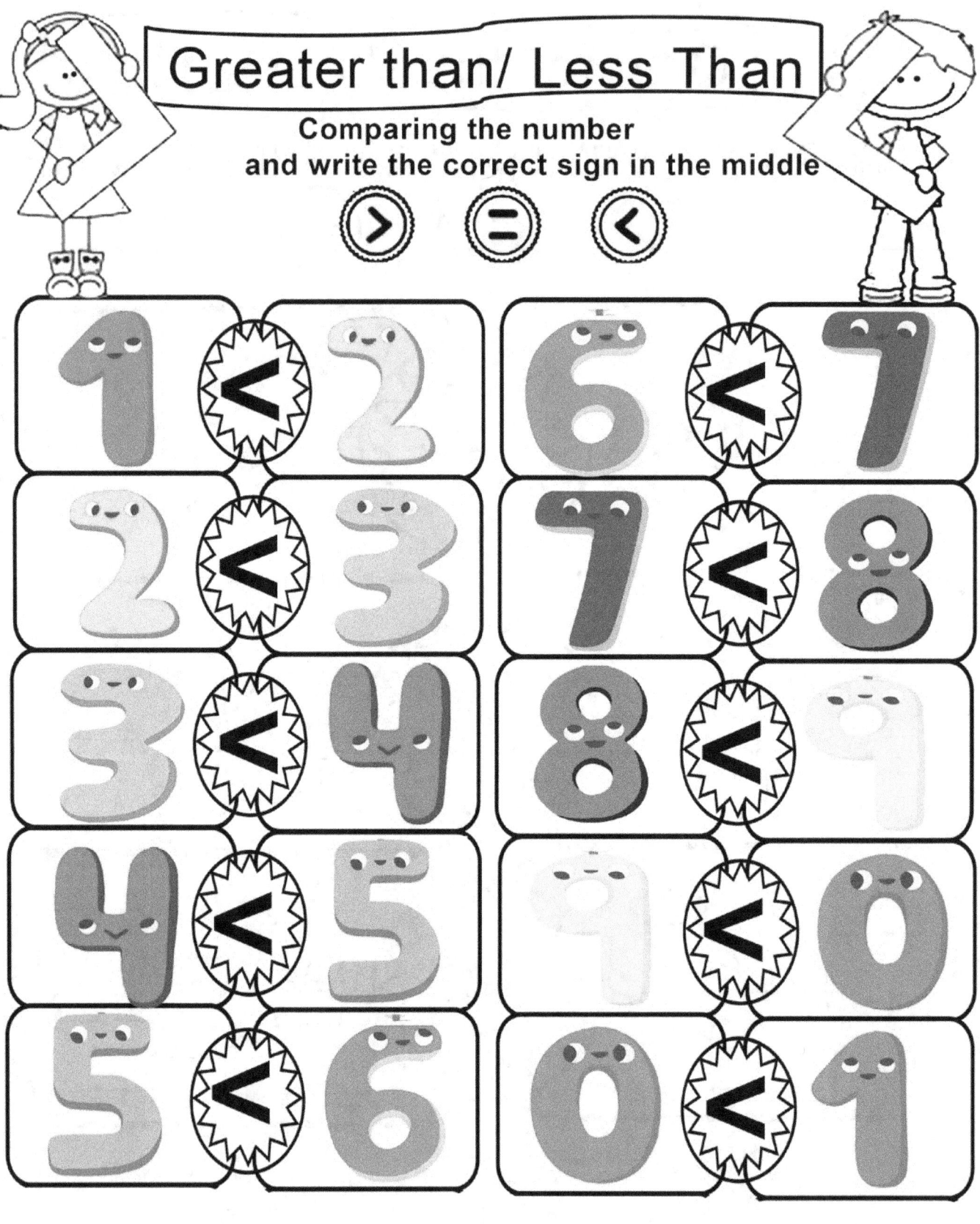

Greater than/ Less Than

Comparing the number and write the correct sign in the middle

> = <

2 > 1	7 > 6
3 > 2	8 > 7
4 > 3	9 > 8
5 > 4	1 > 0
6 > 5	1 > 0

Greater than/ Less Than

Comparing the number and write the correct sign in the middle

> = <

Your Score is

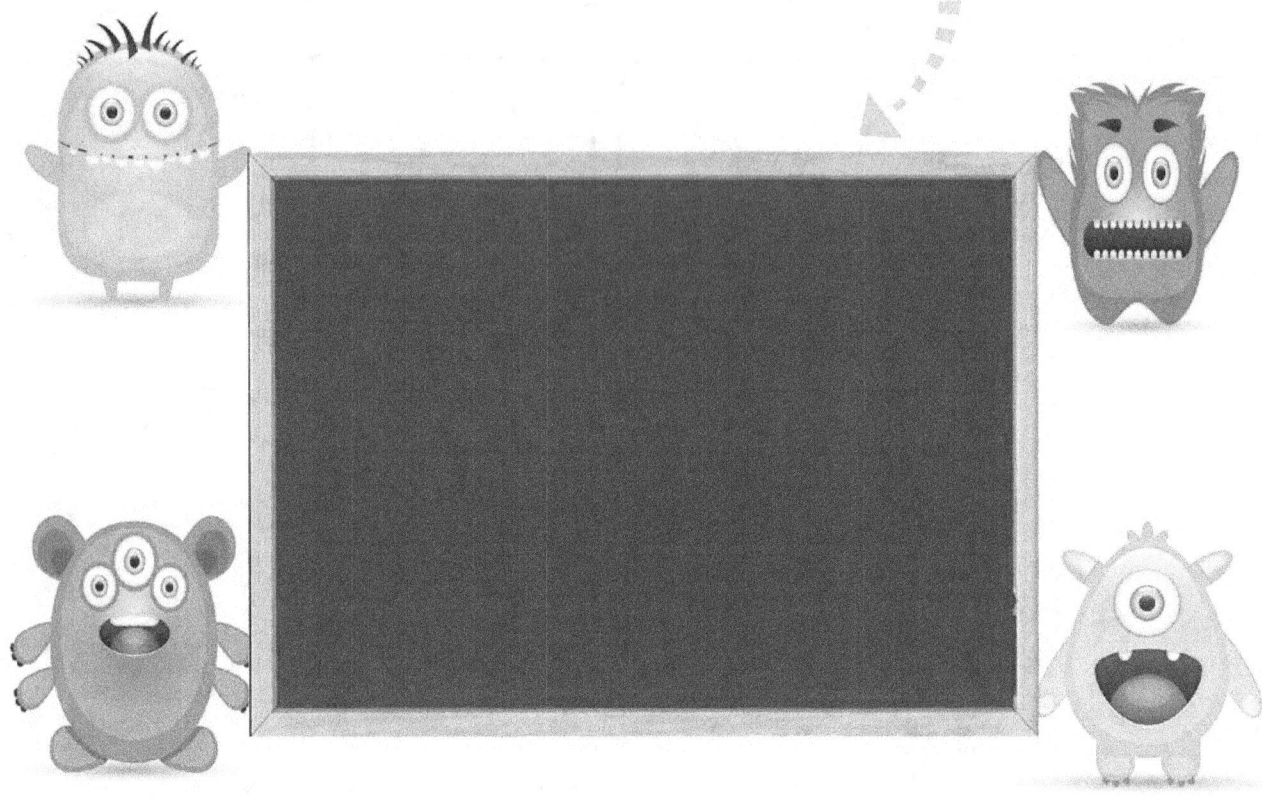

Note

Basic Addition

Learning Math

Count, Add and Write the correct number in the BOX

Count, Add and Write the correct number in the BOX

Count, Add and Write the correct number in the BOX

Count, Add and Write the correct number in the BOX

Count, Add and Write the correct number in the BOX

Count, Add and Write the correct number in the BOX

Count, Add and Write the correct number in the BOX

Count, Add and Write the correct number in the BOX

Count, Add and Write the correct number in the BOX

Count, Add and Write the correct number in the BOX

Basic Addition

Adding Muffin
Count and add

1 + 2 = 3

1 + 1 = 2

2 + 2 = 4

Adding Donuts
Count and add

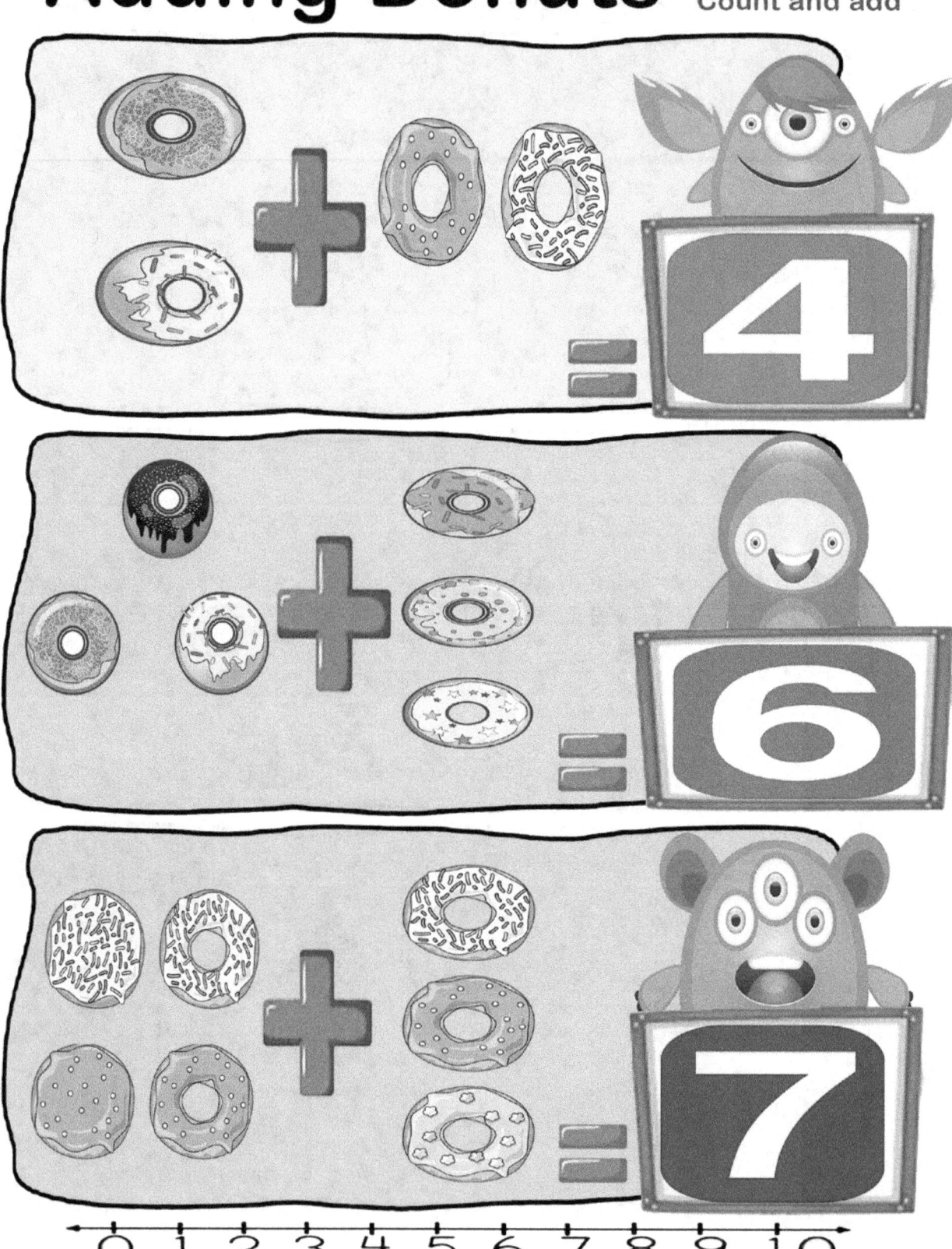

Adding Cupcakes
Count and add

Adding Muffin
Count and add

Adding Ice Cream
Count and add

Adding Macaroon

Count and add

Adding Ice Cream
Count and add

Adding Chocolate

Count and add

Adding Candy

Count and add

Adding Candy

Count and add

Your Score is

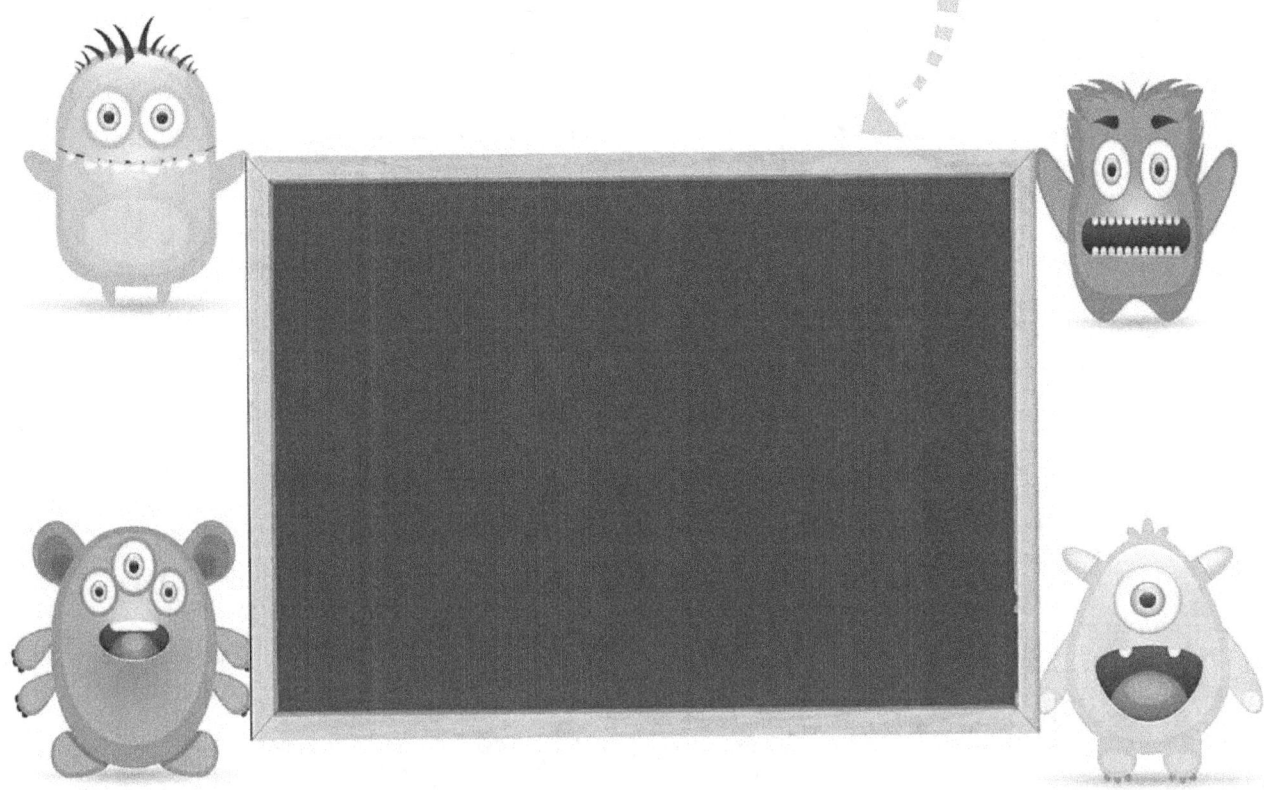

Note

www.ingramcontent.com/pod-product-compliance
Lightning Source LLC
Chambersburg PA
CBHW062333220526
45469CB00008B/2691